Sensor Circuits and Switching for Stringed Instruments

Donald L. Baker

Sensor Circuits and Switching for Stringed Instruments

Humbucking Pairs, Triples, Quads and Beyond

 Springer

Donald L. Baker
Tulsa, OK, USA

ISBN 978-3-030-23126-2   ISBN 978-3-030-23124-8   (eBook)
https://doi.org/10.1007/978-3-030-23124-8

This Springer imprint is published by the registered company Springer Nature Switzerland AG
The registered company address is: Gewerbestrasse 11, 6330 Cham, Switzerland

Preface

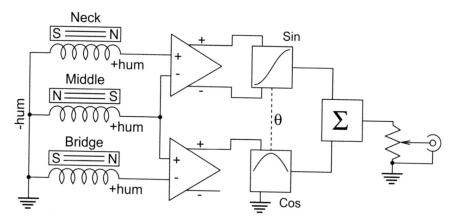

I created this circuit diagram specifically for the cover art; it appears nowhere else in the text. It neatly summarizes three innovations that this book describes:

1. A simplified switching system that with simple rules produces a reduced number of all humbucking circuits from any odd or even number (greater than 1) of matched single-coil pickups (or humbucker coils used as such)
2. A single-coil pickup with a reversible magnet, which allows 2, 4, 8, 16, and up overlapping but different tonal characters of all humbucking tones, for 2, 3, 4, 5, and up matched single-coil pickups
3. A means to combine matched single-coil pickup signals with variable gains to produce not just the tones of all possible switched humbucking circuits but all the tones in between.

But of course, you don't believe that yet. Not until you see and understand the circuits and math.

Tulsa, OK, USA
May 28, 2019

Don Baker

Acknowledgments

I would like to thank Mr. Charles Glaser of Springer Nature for the opportunity to offer this book for publication; Ms. Brinda Megasyamalan and her team, including Ms. Kiruthika Kumar, for their help in setting this work up for publication; and the Massachusetts Institute of Technology for teaching me electronics and linear vector algebra all those years ago. Several software packages were critical to the work: Tina-TI, Version 9.3.50.40 SF-TI, Copyright 1993–2012, DesignSoft, Inc.; Maple 5 Release 4, Ver 4.00c, Copyright 1981–1996, Waterloo Maple Inc.; PhotoShop 7, ver 7.01, Copyright 1990–2002, Adobe Systems, Inc.; Microsoft Excel 2002, SP3, Copyright Microsoft Corp. 1985–2001; Simple Audio Spectrum Analyzer v3.9, Copyright W.A. Steer 2001–2016; Maxwell Version 3.1.04, Copyright 1984–2002, Ansoft Corp.; and PDF24 Creator, Ver. 8.2.2, www.pdf24.org. I would also like to thank Mr. Jim Ziegler of The Music Store and The Zigs in Tulsa, Oklahoma, USA, for kindly playing my prototype guitars and for giving comments and Woodcraft for Tulsa for when I couldn't stand the computer anymore and had to get up and make something with my hands.

Any math errors in this book are definitely mine. I'm good, but not perfect.

—Don Baker, Ph.D., Retired, Tulsa, OK, May 2019

Contents

Chapter 1
Introduction and Short Previews of Coming Chapters

1.1 Some General Comments on Prior Art

The object of the entire book is to produce more than one patented switching and control system that will produce all-humbucking tones from stringed instruments using matched electric-magnetic pickups, eventually leading to a system where the musician can change monotonically from bright to warm tones and back, without ever needing to know which pickups were used in what combinations. In some cases, this may be extended to other sensors and to non-stringed instruments.

For half a century or more, most of the electric guitars sold have just two kinds of electromagnetic pickup setup, with a single kind of switching system for each: the dual-humbucker guitar with a 3-way switch, and the 3-coil guitar with a 5-way switch. As many guitar legends have demonstrated over the years, they work very well. One might think that in all this time, all the answers have been found, electric guitars have been engineered to a fare-thee-well, and nothing more remains to be said.

Not exactly.

This book presents a number of extended and new approaches to developing, constructing, and determining the maximum number of possible series-parallel guitar pickup circuits. Ironically, most of this uses math and engineering that was available to undergraduate electrical engineering (EE) majors in the 1960s, if not long before. It seems that in this field at least, no one had yet bothered to develop it. Other fields and disciplines in EE or math may have, but if so, it has not transferred to electric guitars, pianos, and other instruments. As a result, patents have been granted for pickup switching systems which produce a huge number of duplicate circuits, even circuits with no output, leaving it to the guitarist to figure out which ones are useful. Few if any of these have succeeded widely, or at all, in the marketplace, effectively ceding it to 3-way dual-humbucker and 5-way 3-coil guitars.

Yet, no matter how good an idea may be, there is also a decades-long history of inertia in the guitar and piano (stringed instrument) industry, with a "not invented here" prejudice. Most inventions are incremental, making small tweaks to existing

© Springer Nature Switzerland AG 2020
D. L. Baker, *Sensor Circuits and Switching for Stringed Instruments*,
https://doi.org/10.1007/978-3-030-23124-8_1

art. One cannot expect a well-founded industry to believe that something entirely new under the sun has been developed that will make major contributions to extending the versatility of electric stringed instruments. Such things cannot be accepted until they are conclusively demonstrated, if not by example, then at least by repeatable and irrefutable math and engineering.

There is also the uncertainty of actually getting a patent. The Non-Provisional Patent Application (NPPA), from which the first few chapters arise, began with 3 independent Claims and 24 other Claims dependent upon them. They were not easy to write and were amended several times. As Chap. 2 demonstrates, the method of creating series-parallel circuits is simple, if difficult for larger numbers. It produced duplicate circuits for those with more than three sensors, which were eliminated through visual inspection, i.e., "which of these things looks like another." One might expect that a computer algorithm can be written to do it, but this author lost the ability to do higher math and computer programming to certain medications. So it will be a job for someone else.

Here follows abbreviated descriptions of methods developed to construct series-parallel pickup circuits and switching systems. In each, the switching of pickups may be either electromechanical or digitally controlled, and the signals may be amplified, but the signal path is pure analog. Nothing digital gets between the output to an amplifier and the magic of fingers on strings.

The last chapter eliminates almost all switching in favor of variable gains, which can use either analog or digital controls, and will duplicate not only all the humbucking tones of the series-parallel circuits presented here, but also all the continuous humbucking tones in between. As an added bonus, the humbucking tones of $J > 1$ number of pickups can be produced by $J - 2$ number of variable gain controls. Thus one control can run through all the humbucking tones of an electric guitar with three matched single-coil pickups, some of which, the humbucking triple-coil circuits, have possibly never been heard before.

The object of the entire book is to produce more than one patented switching and control system that will produce all-humbucking tones from stringed instruments using matched electric pickups, eventually leading to a system where the musician can change monotonically from bright to warm tones and back, without ever needing to know which pickups were used in what combinations. In some cases, this might be extended to non-stringed instruments.

1.2   Methods of Creating Pickup Circuits

This presents short descriptions of the methods developed in the following chapters, with some of the results.

1.2.1   Method 1 (Chap. 2)

The first takes a simple method of combining sensors (in this case single-coil electromagnetic pickups, but it could be others, like piezoelectric) together in series and parallel to find the maximum number of unique series-parallel circuits for a given number of pickups, as shown in Table 1.1 below.

Table 1.1 Total numbers of all possibly unique tones from K sensors taken J at a time, including reversing the terminal connections of individual sensors in a circuit

J	1	2	3	4	5	6	
J_V	1	2	4	10	24	72	
J_T	1	2	8	58	502	7219	
N_{SGN}	1	2	4	8	16	32	
$J_T * N_{SGN}$	1	4	32	464	8032	231,008	
K							K_T
1	1						1
2	2	4					6
3	3	12	32				47
4	4	24	128	464			620
5	5	40	320	2320	8032		10,717
6	6	60	640	6960	48,192	231,008	286,866
7	7	84	1120	16,240	168,672	1,617,056	1,803,179
8	8	112	1792	32,480	449,792	6,468,224	6,952,408
9	9	144	2688	58,464	1,012,032	19,404,672	20,478,009
10	10	180	3840	97,440	2,024,064	48,511,680	50,637,214

J is the number of sensors in a circuit. The method for constructing all possible series-parallel combinations is relatively simple. Start with one sensor, and add another to it in series and parallel to obtain $J_V = 2$ versions of $J = 2$ circuit topologies. Then take those two and add one in series and parallel with each to create $J_V = 4$ versions of $J = 3$ circuit topologies. And so on.

Now, the circuits can be changed by exchanging the positions of pickups in the circuit. J_T is the total number of different circuit topologies for J sensors, including the number of ways that individual sensors can be moved to different locations in each circuit topology. Then the terminals of individual sensors in a circuit can be switched to produce phase changes. N_{SGN} is the total number of possible unique sensor terminal reversals; $N_{SGN} = 2^{J-1}$. $J_T * N_{SGN}$ is the total number of ways that J sensors can be connected together in circuit topologies of size J.

If one has $K \geq J$ sensors total, then $J_T * N_{SGN}$ must be multiplied by the combinations of K things taken J at a time to find the total number of different circuit topologies of size J. Then K_T is the total number of different circuits that can be constructed from K or less sensors. Note that only a tiny percentage of these can be humbucking circuits, and then only if the (electromagnetic) pickups are matched to have exactly the same response to external hum.

Note that Table 1.1 cites "possibly unique tones." Most of the signals and tones produced by such large numbers of distinct circuit topologies will be clustered at the warm end. The brightest tones tend to be those from circuits that produce out-of-phase (or contra-phase) effects, from pickups with signals that conflict in phase.

They can sound thin. The warmest tones tend to come from circuits with larger number of J out of K pickups, spread over the space from the neck to the bridge. Frequently, particularly at the warm end, tones from different circuit topologies will be so close together as to be virtually indistinguishable. Thus, depending on one's definition of "distinct tones" there will rarely be as many distinct tones as there are distinct circuits, especially for larger numbers of pickups closer together.

1.2.2 Method 2 (Chap. 3)

The second method inserts a dual-coil humbucker in each sensor position of each circuit topology found by the first method. It then connects the humbucker into that position with the humbucker coils connected either in series or parallel, multiplying the number of possibilities for JJ number of humbuckers by $N_{SP} = 2^{JJ}$. In Table 1.2 below, KK = the number of available humbuckers; JJ = the number of those KK humbuckers used in a circuit topology of JJ size; JJ_T = number of circuit topologies of JJ size, including all the different positions that humbuckers can take within the circuits to produce different signals; $N_{SGN} = 2^{JJ-1}$, the number of unique ways JJ number of humbuckers can be reversed in a circuit topology; $N_{SP} = 2^{JJ}$, the number of internal series-parallel connections of coils in each of JJ humbuckers; and KK_T is the number of total possible unique circuits that can be constructed with KK number of humbuckers or less.

Table 1.2 Total number of possibly unique tones from KK humbuckers taken JJ at a time, with sums KK_T across all possible JJ $\leq$ KK

$N_{SGN} * N_{SP} * JJ_T$	2	16	256	7424	257,024	
JJ_T	1	2	8	58	502	
N_{SP}	2	4	8	16	32	
N_{SGN}	1	2	4	8	16	
JJ	1	2	3	4	5	
KK						KK_T
1	2					2
2	4	16				20
3	6	48	256			310
4	8	96	1024	7424		8552
5	10	160	2560	37,120	257,024	296,874

Table 1.2 shows that instead of just 3 ways, there are 20 ways to connect together the coils of 2 humbuckers into humbucking circuits.

1.2.2.1   An Experiment with 2 Humbuckers

Fig. 1.1 Prototype dual-mini-humbucker guitar with 20-way switching system

Figure 1.1 shows a modified guitar with 2 mini-humbuckers, a 4P6T switch, labeled A–F, and a 4PDT switch for each humbucker that switches its coils between series and parallel connections. The 4P6T switch produces the sequence from A to F, respectively, N + B, N‖B, N, B, (−N) + B, and (−N)‖B, where N refers to the neck humbucker, B refers to the bridge humbucker, "+" means a series connection of the 2 humbuckers together, "‖" means the humbuckers connected in parallel with each other, and (−N) means that the terminals of the neck humbucker are reversed in phase compared to the bridge humbucker. This produces 24 switch combinations, of which 4 are duplicate circuits, leaving 20 distinct circuits.

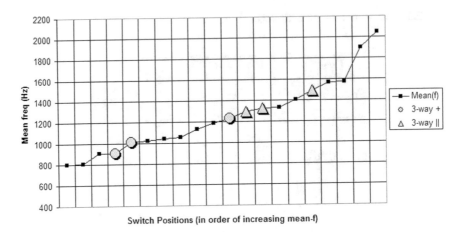

Fig. 1.2 Mean frequencies of six strummed strings in the 20-way switching system, in order of increasing mean frequency

Figure 1.2 shows mean frequencies of the outputs of the switch positions of the 20-way switching system in Fig. 1.1, taken for 6 strings strummed 5 times, measured with an FFT program at 44.1 kHz sampling rate, 4096 samples per Hann window, with a frequency resolution of about 10.8 Hz. The triangles show the equivalent 3-way switch positions for humbuckers with internal coils connected in parallel. The circles show the same 3-way switch positions for humbuckers with series-connected internal coils. Of the 20 possible tones, 3 mean frequencies are closer together than 8 Hz, and 4 are closer together than 10 Hz, leaving 16 or 17 possibly unique tones, which are separated in mean frequency by 15 Hz or more. The mean frequencies extend over a range of about 1.36 octaves or the equivalent of about 16.3 frets.

1.2.3   Method 3 (Chap. 4)

This depends upon the use of matched single-coil pickups, which all have the same response to external hum, not a common consideration in current pickup manufacture. Like humbuckers, their circuits are constructed by doubling the pickups in Method 1, with series and parallel pairs, also known as humbucking pairs. If two match-coil pickups have different magnetic poles up toward the strings, they can only be connected together in-phase to be humbucking. If 2 such pickups have the same pole up, they can only be connected together contra-phase to be humbucking. Because of this, humbucking pairs must reverse connections together, as a whole. The humbucking pairs presented here are connected together in pairs, quads, and hextets. Octets and up are left as an exercise for the reader.

This chapter distinguishes between loaded and no-load tones. Series and parallel circuits are often only different in tone because of the loading by volume pots and tone controls causing high-frequency roll-offs (reductions), more for series circuits than parallel circuits. If only the open-circuit output, or no-load, equations are considered, series and parallel circuits often have the same relative contributions from the same pickups, differing only in amplitude of signal and the equivalent output inductance and resistance. So when the outputs are fed into high-impedance input preamps or amplifiers, particularly with integrated circuits, this reduces the number of different actual tones.

Tables 1.3 and 1.4 show the numbers of loaded and no-load tones, respectively.

Table 1.3 Numbers and sums for loaded tones for humbucking pairs, quads, hextets, and octets, for numbers of matched single-coil pickups from 2 to 8. The inner numbers are NLT times (K pickups taken J at a time). The sums show how many potentially unique loaded tones can be had K pickups using pairs, quads, and hextets. Adapted and corrected from Math 31, Baker (2017a)

$J =$	2	4	6	
$N_{LT} =$	2	33	2200	
K	Pairs	Quads	Hextets	Sums
2	2			2
3	6			6
4	12	33		45
5	20	165		185
6	30	495	2200	2725
7	42	1155	15,400	16,597
8	56	2310	61,600	63,966

Table 1.4 Numbers and sums for no-load tones for humbucking pairs, quads, and hextets, for numbers of matched single-coil pickups from 2 to 8

$J =$	2	4	6	
$N_{NL} =$	1	27	1000	
K	Pairs	Quads	Hextets	Sums
2	1			1
3	3			3
4	6	27		33
5	10	135		145
6	15	405	1000	1420
7	21	945	7000	7966
8	28	1890	28,000	29,918

Chapter 5 discusses the limits of mechanical switches to produce all of these circuits.

Chapter 6 discusses the use of a digitally controlled analog cross-point switch to more efficiently and effectively produce all of these combinations, and defines a micro-controller system architecture for that purpose.

1.2.4   Method 4 (Chap. 7)

One can produce even more tones from the same circuits, constructed from matched single-coil pickups by simply reversing the magnets individually. (Note to Patent Examiner: Reversing the magnets in dual-coil humbuckers, which have only one magnet, merely reverses the string signal polarity of the humbucker, an operation which could be more easily done by reversing the leads.) For each set of J pickups, there are 2^{J-1} number of pole configurations, 2, 4, 8, and 16 for $J = 2, 3, 4$, and 5.

Reversing the magnets does not affect humbucking. Humbucking depends only upon the coil circuit connections and the pickups having equivalent magnetic paths and coil responses to external hum. Reversing magnets affects which coils will be in-phase or contra-phase with each other, and the relative number of in-phase and contra-phase humbucking circuits. When the coils are fixed in place under the strings, as usual, changing which pairs are in-phase affects the tone.

But there are more than just humbucking pairs, quads, and hexes. Baker developed a humbucking triple-coil circuit (2019a) and simplified humbucking circuits for odd numbers of coils from 5 up (2018b). The introduction of no-load tones requires that all of the series-parallel circuits developed in Chap. 2 be checked for humbucking properties. It turns out that there is 1 no-load tone circuit for 3 pickups, 3 no-load tone circuits for 4 pickups, and 8 no-load tone circuits for 5 pickups. The number of no-load humbucking circuits for 6 matched pickups are presented in Table 11.1.

New rules for constructing tone circuits from no-load circuit output equations are discussed. In this way, for circuits of 3 matched pickups and up, new humbucking circuits can be found that Method 3 did not. These include new rules for pickup and phase combinations, which were not necessarily back-applied to previous chapters. The possible pole configurations can be different, but not entirely unique, because they must share some tones. Therefore the output selections for the number of tones using pickups with reversible magnets is more than the total number of unique tone circuits.

Tables 1.5, 1.6, 1.7, and 1.8 summarize the results. See Chap. 7 for more details.

Table 1.5 Humbucking tones for $J = 2$ to 5, $K = 2$ to 8, where J is the number of pickups in the humbucking circuit, HB Expr is the number of humbucking output expressions for circuits of J matched pickups, hum Expr is the number of non-humbucking output expressions for J pickups, HB Ckts is the number of pickup combinations for all the humbucking output expressions, K is the total number of pickups available to be switched, sum of tones is the number of pickup combinations for all $J \leq K$

$J =$	2	3	4	5	
HB Expr	1	1	3	8	
hum Expr	0	1	1	3	
HB Ckts	1	3	19	365	Sum of tones
K					
2	1				1
3	3	3			6
4	6	12	19		37
5	10	30	95	365	500
6	15	60	285	2190	2550

Table 1.6 Cumulative sums of tonal circuits for $J \leq K$ numbers of matched pickups with reversible magnets, over all the 2^{J-1} pole configurations, where N_K is the number of possibilities for K matched pickups, and the rows below show the combinations for $J \leq K$

$J =$	2	3	4	5	
$N_K =$	2	12	52	2720	
K					Sums
2	2				2
3	6	12			18
4	12	48	52		112
5	20	120	260	2720	3120
6	30	240	780	16,320	17,370

Chapter 7 also discloses embodiments for single-coil pickups with reversible magnets.

1.2.5 Method 5 (Chap. 8)

The switching of humbucking pickup circuits can be vastly simplified by using only a certain type of series-parallel circuits, governed by a few simple rules, in exchange for fewer choices, but still much greater than generally available today. It involves connecting all terminals of the matched single-coil pickups with the same hum polarity together at a common connection point (Baker, 2018b), preferably at the outer windings to improve shielding. Either the common connection point can be grounded, or the designated low terminal of the switching output, but not both. At least one pickup must be connected to each output terminal. Tables 1.7 and 1.8 summarize the resulting choices for a single pole output configuration.

Table 1.7 Numbers of circuits for K pickups taken J at a time in a common connection point switching circuit, for one pole configuration

$J =$	2	3	4	5	6	7	8	9	10	11	12	Totals
K												
2	1											1
3	3	3										6
4	6	12	7									25
5	10	30	35	15								90
6	15	60	105	90	31							301
7	21	105	245	315	217	63						966
8	28	168	490	840	868	504	127					3025
9	36	252	882	1890	2604	2268	1143	255				9330
10	45	360	1470	3780	6510	7560	5715	2550	511			28,501
11	55	495	2310	6930	14,322	20,790	20,955	14,025	5621	1023		86,526
12	66	660	3465	11,880	28,644	49,896	62,865	56,100	33,726	12,276	2047	261,625

Table 1.8 Total number of tone circuits for K pickups taken J at a time in a common connection point switching circuit, over all 2^{K-1} pole configurations

$J =$	2	3	4	5	6	Totals
$2^{J-1} =$	2	4	8	16	32	
K						
2	2					2
3	6	12				18
4	12	48	56			116
5	20	120	280	240		660
6	30	240	840	1440	992	3542

Chapter 9 discloses an experiment with two mini-humbuckers using this method, treating the dual coils as 4 separate matched coils, producing 25 different humbucking outputs from pair, triple, and quad coil circuits. Figure 1.3 shows a plot of the results, which look very similar to the previous dual-humbucker series-parallel experiment, in Fig. 1.2, using the same guitar and pickups.

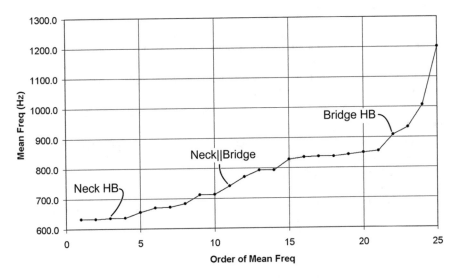

Fig. 1.3 Shows the results of humbucking circuits from Table 9.3 for mean frequency versus frequency order. It highlights the equivalent 3-way switch results, the neck humbucker (Neck HB) at the 3rd spot, 636.1 Hz, the neck and bridge humbuckers in parallel (Neck||Bridge) at the 11th spot, 741.4 Hz, and the bridge humbucker (Bridge HB) at the 22nd spot, 907.5 Hz. It shows a number of frequencies bunched closely together, at 632.9 to 639.4 Hz, 669.8 and 672.2 Hz, 712.6 and 713.5 Hz, 792.8 and 792.9 Hz, and from 835.0 to 837.2 Hz

The tones for this experiment, as expressed by the mean frequency of their spectra, clearly bunch together in some regions. This chapter also explores the ordering of tones, according to the mean frequency of their spectra, in equal equivalent fret steps.

Chapter 10 examines several embodiments of switching systems for this method, including mechanicals switching for 3 and 4 matched pickups, and expandable digital switching for more pickups, which requires only one inexpensive solid-state 1-pole-3-throw switch per pickup, plus a few other simple digital-analog switches for other functions. Digital switching can be used with or without a micro-controller (uC) system. But using a uC allows digital samples of switch signals, calculation of FFT spectra, and ordering of the tones for selection by a user interface.

1.2.6 Method 6 (Chap. 11)

There is a radically different way to look at combining pickups signals to generate different humbucking tones that does not involve switching. This method can use either active pickups, where a low-gain preamplifier isolates a single pickup, or humbucking pairs of pickups feeding into fully differential amplifiers. Both approaches use variable gains to control the mix of humbucking signal, then combine the signals in a summer.

This approach uses linear vector math, reducing J pickups into $J - 1$ humbucking pairs, which overlap in the signal path, i.e., $s(A–B) + u(B–C) + v(C–D) + \ldots$, where A to D are placeholders, used both for pickup designations and hum voltages, to be replaced by positive string signals for North-up pickups and inverted negative string signals for South-up pickups. The $J - 1$ coefficients, $s, u, \ldots$ form an n-dimensional vector SUV-space, which the chapter explores. When they are replaced by combinations of dual-gang sine/cosine analog or digital pots, the number of controls can be $J - 2$. So a 3-coil guitar only needs one control to navigate the different tones in the SUV-space.

It turns out that every humbucking circuit output equation can be transformed to an SUV-space equation. This means that for J pickups, the spectra of $J - 1$ humbucking pairs can be used to construct the spectra of all the points in the SUV-space, giving results for tones and relative amplitudes.

The chapter explores construction of psuedo-sine/cosine pots from linear pots, and the accuracy of using digital linear pots to approximate sine and cosine pots. It explores an algorithm for constructing sines and cosines from 32-bit floating point 4-function arithmetic, plus square root, which calculate sine and cosine, accurate from ± 0.06 to ± 0.007 to ± 0.000015, depending on the complexity chosen, which can then be used in calculating FFTs as well as SUV coefficients.

The chapter also adapts the uC system presented in Chaps. 6 and 10 to this application.

1.3   There Are Several Good Reasons for Writing This Book

1. The electric musical instrument industry has a lot of inertia. Much of it, not all, is based upon reproducing the instruments that various musical heroes used, so that the budding hero can sound just like them. In electric guitars, those designs go back 50 years or more and are resistant to change, partly because of the not-invented-here syndrome, favored by companies which prefer to own the patents they use. So far, most of the technological advances in guitar electronics have been limited to high-end, expensive, and semi-custom guitars. Most of the improvements presented here can be incorporated into mass-market guitars for the price of some batteries and integrated circuits, greatly increasing the range and resolution of available tones, some of which, like humbucking triple-coil and five-coil circuits, may not have ever been heard before.
   Very few are likely to believe this can be done without first seeing the examples and math.
2. The book may be very useful in protecting the intellectual property within it. All of the new work presented here derives from Provisional and Non-Provisional Patent Applications filed with the U.S. Patent and Trademark Office. And a patent is only as good as one's ability to market it to industry and to defend it in court. There is a long history of small inventors losing out to large corporations with well-paid lawyers in deep pockets, even after patenting an invention and spending tens if not hundreds of thousands of dollars on patent lawyers and court filings. There is a story[1] about the inventor of an aortic heart valve (Norred, T. (2000). U.S. Patent 6,482,228), which begat a 1.5 billion dollar a year industry, for which Dr. Norred never got much compensation or recognition. That may be a bit harder to do if one writes the book on it.
3. Pro Se patent applicants (those prosecuting patents without patent lawyers) historically fail at a much greater rate than those represented by lawyers. This author could find only a handful of scholarly articles on the matter, 2 out of 3 blaming the lack of lawyers. None observed that the U.S. Patent and Trademark Office (USPTO) is very resistant, if not outright hostile, to Pro Se inventors, preferring to work instead with lawyers. Never mind the fact that this author's monthly Social Security income would not pay a patent lawyer for 8 billable hours. In this author's experience, indications of USPTO disapproval include:

   (a) Outright insistence that a lawyer should be used, because a Pro Se applicant allegedly causes the patent examiner "too much trouble." Among other things, that patent examiner seemed to take offense at the very idea of writing patent applications as clear engineering tutorials, intended for later publication as a book or technical papers.
   (b) Patent examiners using derogatory language in their first evaluations of claims, including "replete with errors," and "replete with indefinite

---

[1]Eden, S. (2016, July/August). The Greatest American Invention. *Popular Mechanics,* 92–99. Retrieved from https://www.popularmechanics.com/technology/a21181/greatest-american-invention/

language," without necessarily providing clear descriptions or alternatives. In one case, the examiner complained in conference that it took him "a whole hour" to read the first claim. Then, when the applicant offered to take a non-final rejection to allow the examiner more time to review the work, and the applicant to review and rewrite the claims, the examiner claimed that he did it, so magnanimously, "as a courtesy to the Applicant."

(c) An outright lie about whether the applicant had properly filed a mailing address, requiring multiple and arbitrary rounds of paperwork to resolve the matter.

(d) Demands for "corrections" to paperwork and applications, couched in multiple and obscure references to U.S. Code (USC), the Code of Federal Regulations (CFR) and the Manual for Patent Examination Procedure (MPEP), without the use of plain language, as mandated by the Federal Paperwork Control and Plain Writing Acts.

(e) Several cases of examiners and staff trying to get an applicant to insert fatal language in an application, which would have made it easier to dismiss.

(f) One patent examiner reduced 20+ claims to a two-word phrase, encouraging the applicant to salt it through the amended claims. Whereupon the examiner used it to attack the claims, stating among other things that it was "obvious" because he put his own phrase into a term search and got 14,000 hits. At the time, "flat earth" got 11,000,000 hits on Google.

(g) Specious arguments by patent examiners, misrepresenting and even falsifying prior art, in order to reject claims. One examiner falsely claimed a passive dual-coil humbucking pickup had an active, powered solid-state preamplifier embedded in its structure. The presence or absence of which has no practical relevance to the invention, a reversible magnet on a passive single-coil pickup. And yet, as of this writing, this is not yet a firing offense at the USPTO.

(h) The same patent examiner, stating that prior art anticipated varying the shape and properties of magnets in electromagnetic pickup coils, cited in his reference list a patent (allowed by a previous examiner), where the various constructions of the pickup magnets tended to short out the field with materials acting like keeper bars. One embodiment of a pickup has no magnet. That patent's examiner did not know the difference. This allows prior bunkum to "anticipate" and disallow anything that does work.

(i) The practice of the Department of Commerce Office of Inspector General in referring, as not worthy of its attention, complaints against the Patent Office back to the Patent Office, which then had someone call to express "regrets" for one's "frustration," and to say that one is "being heard." Followed by the next patent examiner making false statements about prior art to reject claims. And the DoC OIG refusing to accept any more complaints. The word "complicity" comes to mind.

See   https://www.researchgate.net/project/NPPA-15-616-396-China-is-not-the-most-clear-and-present-danger

And   https://www.researchgate.net/project/US-Patent-App-15-917-389-Single-Coil-Pickup-with-Reversible-Magnet-Pole-Sensor

That makes this book a matter of self-defense. If the USPTO refuses to grant patents on false and specious grounds, then this book may stand as testimony against those allegations, perhaps even in Federal Courts.

It also illustrates the fallacy of collegiality, which seems to encourage this kind of behavior by suppressing negative and inconvenient information. And it also raises the question of entrenched discrimination, of keeping out undesirables. For poor and colored are the very people who cannot afford lawyers. Witness the historically biased treatment of black and other colored Americans in getting USDA farm loans and Federal home loans. And just like home ownership and farm operation, intellectual property forms a source of wealth and relief from poverty.

Those of us who have lived on disability also feel the sting. After working for nearly a decade and a half to get back to taxpaying work and productivity, getting pushed back down into dependence upon social programs, by the very same Conservative Administration that condemns them, cannot be excused or forgiven.

Patent lawyers, of course, are quite willing to show collegial courtesy in putting up with this behavior and excess paperwork from the USPTO. It increases their billable hours, and biases the USPTO against Pro Se inventors, who are more likely to lose their tempers at such treatment. Not to mention that obsessive-compulsive paperwork creates a huge backlog in unprocessed patents, further increasing billable hours.

References

Baker, D. L. (2017a). Humbucking switching arrangements and methods for stringed instrument pickups, US Patent Application 15/616,396, filed 7 June 2017, published as US-2018-0357993-A1, 13 Dec 2018, granted as Patent US10,217,450, 26 Feb 2019. Retrieved from https://www.researchgate.net/publication/335727402_Humbucking_switching_arrangements_and_methods_for_stringed_instrument_pickups_-_NPPA_15616396

Baker, D. L. (2018b). Means and methods for switching odd and even numbers of matched pickups to produce all humbucking tones, US Patent Application 16/139,027, 22 Sep 2018, published as US-2019-0057678-A1, Feb 21, 2019, granted as U.S. Patent 10,380,986, 08/13/2019. Retrieved from https://www.researchgate.net/publication/335728060_NPPA-16-139027-odd-even-HB-pu-ckts-2018-06-22

Baker, D. L. (2019a). Humbucking switching arrangements and methods for stringed instrument pickups, US Patent 10,217,450, filed 7 June 2017, granted 26 Feb 2019. Retrieved from https://patents.google.com/patent/US10217450B2/

Eden, S. (2016, July/August). The greatest American invention. *Popular Mechanics*, 92–99. Retrieved from https://www.popularmechanics.com/technology/a21181/greatest-american-invention/

Chapter 2
Series-Parallel Circuit Topologies of Single Sensors

2.1 Some Prior Art

This invention relates to the electronic design of stringed instruments, including guitars, sitars, basses, viols, and in some cases pianos, including the areas of the control of the timbre of electromagnetic and other transducers by means of combinatorial switching and analog signal processing. Some of the principles will also apply to combinations of other vibration sensors, such as microphone and piezoelectric pickups, placed in or on different parts of a musical instrument, stringed or not.

Electromagnetic guitar pickups go back to at least Miessner (1933, US915858), in which ferro-magnetic strings sat between magnetic poles. Fender and Kaufmann (1948, US455575) also patented a pickup with strings passing through it. Morrison (1951, US557754) shows a pickup with a single coil and magnetic poles for each string that looks and works just like modern single-coil pickups. Pickup patents with variations on those basic features include Stich (1975, US916751), Schaller (1985, US4535668), Knapp (1994, US5292998), Lace (1995a, US5389731; 1995b, US5408043), and Damm (2002, US6372976B2). This list is illustrative, not exhaustive; many more guitar pickup patents have been filed.

Most 3-coil guitars have a single-coil pickup at the neck, the middle, and the bridge. They nominally have similar responses to external hum, but the bridge pickup is often hotter to make up for the smaller movement of the strings near the bridge. Moving the standard 5-way switch lever from the bridge to the neck produces outputs for the B, B∥M, M, M∥N, and N pickups, where B, M, and N mean the bridge, middle, and neck pickups, and "∥" means the 2 pickups connected in parallel. Since the middle pickup usually has the one magnetic pole up toward the strings, and the other 2 pickups have the other pole up, the parallel connections are nominally humbucking. The other three switch positions produce outputs that can be affected by external hum.

Single-coil pickups are usually considered "brighter" in tone than dual-coil humbuckers. Because dual-coil humbuckers usually have their internal coils

© Springer Nature Switzerland AG 2020
D. L. Baker, *Sensor Circuits and Switching for Stringed Instruments*,
https://doi.org/10.1007/978-3-030-23124-8_2

connected in series, the overall inductance of the combination tends to be higher than a single-coil pickup. This means that for the same external load, presented by the tone pot, volume pot, guitar cable, and amplifier input, the inevitable high-frequency roll-off (decreasing output for higher frequency) will be audibly lower.

A great many patents have been filed and granted regarding guitar pickup switching systems to connect various combinations of single-coil pickups to an output with the intent of generating a wide and unique range of usable tones. They include, as examples, Fender (1966, US3290424), Simon (1979, US4175462), Peavey (1981, US4305320), Starr (1987, US4711149), Saunders (1989, US4817486), Wolstein (1992, US5136919), Riboloff (1994, US5311806), Thompson (1998, US5763808), Furst and Boxer (2001, US6316713 B1), Olvera and Olvera (2004, US6781050 B2), Krozack et al. (2005, US2005/0150364A1), Wnorowski (2006, US6998529 B2), Jacob (2009, US2009/0308233 A1), and Hamilton (2011, US7999171 B1). Most of these can be downloaded in PDF format from patents.google.com.

2.2 Why Did They Fail in the Marketplace?

Yet the simple 5-way switch for a 3-coil guitar still dominates in the market. Why? It's cheap, it's simple, it works, and everyone knows how to play it. Consider Fender's (US3,290,424, 1966) patent. It had four single-coil pickups hidden under the pickguard. Each pickup had 2 rows of magnetic poles inside the coil to assure that the field could reach the strings through the pickguard. Each of the four pickups had a 2-pole, 3-thow switch that put the coil either out of the circuit, in the circuit with normal phase, or in the circuit with reversed phase. All coils in the circuit were connected in parallel.

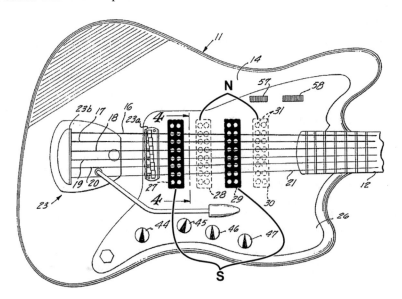

Fig. 2.1 Figure 2 from the 1966 Fender US Patent, 3290424, showing the layout of 4 single-coil pickups

The guitar it represents failed in the marketplace. One anonymous web page claimed that it was considered "too noisy." But that is not all.

Figure 2.1 shows the layout of the 4 pickups, hidden under the pickguard. Figure 2.2, below, shows the 81-way switching system Fender devised for the guitar. Each of switches 44–47 are double-pole-triple-throw (2P3T). Switch 44 is shown connecting the S-up pickup coil 34 to the volume pot 56 in the "normal" position. Switch 45 shows the N-up coil 28 removed from the circuit entirely. And switch 46 shows the other S-up coil, 29, connected to the output with its terminal reversed, so that its signal will be out-of-phase, or contra-phase to the other S-up coil. Switch 47 connects the N-up coil 30 to the output in normal phase.

Note from Fig. 2.2 that these pickups can only be connected together in parallel. Four 3-throw switches means that the number of possible switch combinations $= 3^4 = 81$. One of these combinations has no signal at all, because all the pickups are disconnected from the output. Of the remaining 80 combinations, 40 must have duplicate tones, because each of those circuits has a twin that connects to the output with the opposite phase. But the human ear cannot detect that if there is no other external reference. For any particular frequency, a 180° difference in phase is the equivalent of shifting the signal half a cycle in time. Nor do human ears come with atomic clocks to measure that miniscule time difference. Of the 40 remaining circuit combinations, only the following 9 unique parallel outputs (where a minus sign represents a reversed coil), (N1, S1), (N1, −N2), (N1, S2), (S1, N2), (S1, −S2), (N2, S2), (N1, S1, N2, S2), (N1, S1, −N2, −S2), and (N1, −S1, −N2, S2), can be humbucking, as will be demonstrated later.

It likely failed in the marketplace because it had too many duplicate and non-humbucking choices, with no map to the tones. When the pickup circuit combinations are actually drawn out and analyzed, this is a common failing in many patented switching systems. Too complicated and no map to the useful tones.

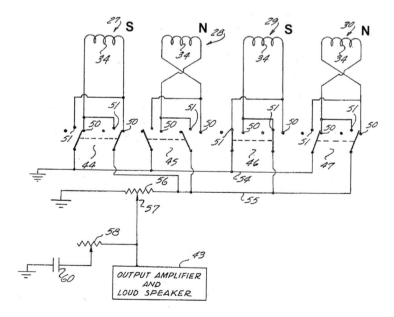

Fig. 2.2 The 81-way switching system for Fender's (US3,290,424, 1966) patent

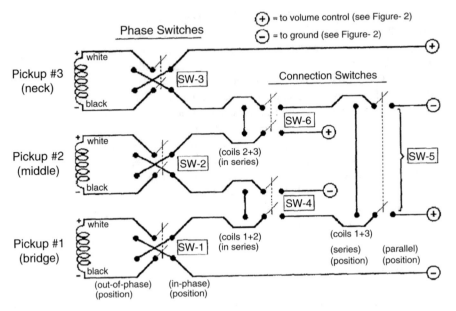

Fig. 2.3 The 216-way switching circuit for Wnorowski (2006, US6998529 B2)

Wnorowski (2006, US6998529 B2) went even farther. For just 3 single-coil pickups, it had 6 switches, as shown in Fig. 2.3. The 3 Phase Switches were all 2P3T, with a center-off position, removing the pickup from the circuit, which had $3^3 = 26$ different switch configurations. The 3 2P2T Connection Switches had $2^3 = 8$ different switch configurations, for a total of $26 * 8 = 216$ different switch configurations. For all these switch configurations, the patent claimed only different 29 different pickup tones. Without regard to phase, the switches could produce the following circuits from pickups (1), (2) and (3): (1), (2), (3), (1 + 2), (1 + 3), (2 + 3), (1 + 2 + 3), (1)‖(2), (1)‖(3), (2)‖(3), (1)‖(2 + 3), (2)‖(1 + 3), (3)‖(1 + 2) and (1)‖ (2)‖(3), where "+" means a series connection and "‖" means a parallel connections.

For J pickups, there are only 2^{J-1} possible unique changes in phase, since the human ear cannot tell the difference if the entire output changes phase by reversing the output terminals. This means for 1 pickup, 1 phase; 2 pickups, 2 phases; and 3 pickups, 4 phases. That means the Wnorowski circuit can produce $3 * 1 + 6 * 2 + 5 * 4 = 35$ different circuits with possibly different tones. Somehow Wnorowski missed 6 of his circuits. Of the remaining 181 switch configurations, about 21 will have no output, and the rest are duplicate circuits.

Not the kind of system a guitarist would care to fiddle with in the middle of a gig. Now if Wnorowski had come up with a switching system that had only those 35 switch configurations, especially if they could have been arranged to switch sequentially in tone from bright to warm, that would have been something. But as we can determine later, many if not most of those 35 configurations would not have been humbucking, and the tones would have likely been bunched at the warm end.

2.3   The Point of the Exercise

Patents like these go at the problem the wrong way. They imagine a switching system before determining just how many different pickup circuits are possible, so that they can't figure out the most efficient way to do the switching. The object is to get the most tones with the least number of switches, preferably with no duplicates or null outputs. With fewer switches and fewer wires connecting them, it's cheaper, more reliable, and less confusing.

The rest of this chapter will systematically construct the maximum number of series-parallel circuit topologies for J number of single-coil pickups, as well as many other types of sensors, which can be directly substituted. When those are constructed, it will show how many ways pickups can be moved from circuit position to circuit position to hopefully change tones, how many ways individual pickups can reverse their terminals to create phase differences, and how many ways one can construct circuits of J number of pickups, $J = 1, 2, 3$, etc., from K number of pickups, where $J \leq K$. We will find out how to get a total of 620 circuits of potentially unique tones out of just 4 single-coil pickups (most of which will not be humbucking). Why "potentially unique"? Because we will find by experiment that tones tend to bunch together at the warm end, and some are so close together as not to count.

In later chapters, we will cover the construction of circuits for dual-coil, and humbucking circuits for single-coil pickups, all matched to respond to hum equally. Eventually, we will do away with switching entirely, producing a continuous range of tones with variable gains. The ultimate object is a system with which the guitarist can switch sequentially or continuously from bight to warm tones and back with a single control, without ever needing to know which pickups are used in what circuits.

But first, we need a glossary of terms to describe different aspects of circuit topology.

2.4   Glossary of Necessary Terms

As you might have guessed, the developments here will not always be easy to follow. It makes it a lot easier to first define some common terms for this kind of circuit topology, and then see how they are used. It helps to keep the math straight. These terms may not be the same as mathematicians may use in other fields of topology, but they will serve here.

Base or basic topology: A collection of one or more sensors all connected in series between two terminals or nodes of a circuit or topology, or alternatively all connected in parallel between two terminals, such that the mere order of connection of sensors in the topology, without changing phases, cannot change the output of the collection in any manner that the human ear or electrical measuring instrument can detect.

Category: The size of a topology, i.e., the number of sensors in a topology, usually designated here by (J) or (M) or a number in parentheses, i.e., (3).

Parallel connection: Two or more two-terminal sensors with one terminal each connected to one circuit node or output terminal, and the other terminal each connected to another circuit node or output terminal.

Phase: The relative reversal of terminals of a sensor or group of sensors in a topology, compared to other sensors in the circuit, such that the human ear can detect a difference.

Series connection: Two or more sensors of two terminals each, connected between 2 output terminals, with one terminal of each sensor connected to the next sensor in a line, which in turn is connected to the next, et cetera, until the last terminal of the last pickup is connected to the other output terminal.

Signs & pairs: The number of potentially unique outputs due to the use of humbucking pairs, for any number of pairs more than 1, $JP \geq 2$. Once used in a patent application, and replaced here in Chap. 4.

Subcategory: A number or sum of numbers, enclosed here in brackets or parentheses, such as $(M1 + M2 + M3)$ or $[4 + 1]$ or $(3 + 2 + 1)$ or $(2 + 1 + 1 + 1)$, indicating a topology of size $M = J$ or category $(M = M1 + \cdots + MN)$, which comprises N number of base topologies, each of size Mi, $i = 1$ to N. The order of Mi number of sensors inside the associated base topology cannot affect the output of the whole, but the reversal of terminal connections of the base topology may.

Topology: The electrical connections of sensors or groups of sensors, particularly two-terminal sensors in series or parallel with respect to each other, such that the output also has two terminals.

Versions: In this context, the number of possible topologies within a subcategory in which exchanging a single sensor with another in the topology will change the output without changing the topology.

2.5 Basic Topologies and Phase

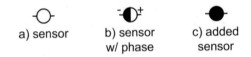

a) sensor   b) sensor   c) added
                   w/ phase   sensor

Fig. 2.4 Symbols for basic sensors of any type. (**a**) sensor, (**b**) sensor w/ phase, (**c**) added sensor

Although we will be working primarily with electromagnetic coil guitar pickups, the symbols in Fig. 2.4 emphasize that in this first chapter the circuit topologies can apply to any kind of sensor with 2 terminals. Figure 2.4a shows the basic symbol for a sensor. Figure 2.4b shows the symbol when we wish to address phase, the negative phase terminal being at the black half, and the positive phase terminal connected to the white half. The "−" and "+" symbols will generally be dropped in use. The blackened circle in Fig. 2.4c will make it easier to understand how sensors are added to construct topologies.

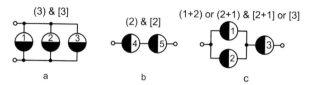

Fig. 2.5 Simple examples of basic and complex topologies, categories and subcategories: (**a**) (3), (**b**) (2) and (**c**) (2+1)

Figure 2.5 shows three simple topologies, of category 3, 2 and 3 (subcategory [2 + 1]). The smaller circles represent the terminals of the entire topology, and the solid dots represent circuit connections. Figure 2.5a shows a basic parallel topology of (3) sensors in parallel, which all happen to be in-phase. Figure 2.5b shows a basic series topology of (2) sensors in series, which all happen to be in-phase. Figure 2.5c shows a basic parallel topology of (2) in series with a basic topology of (1). It can be referred to as a subcategory (2 + 1) or (1 + 2) of category [3], or as a construction of categories [2] and [1], or [2 + 1]. Wait a bit, this is not as redundant as it seems. It has to do with how different subcategories arise from the combinations of categories, which will be demonstrated shortly.

In Fig. 2.5a, if it maintains its phase with respect to the output terminals and the other sensors, sensor 1 can exchange positions with sensors 2 or 3, and it will have no effect on the output signal, including the amplitude, phase, and tone. The same holds true of sensors 4 and 5 in Fig. 2.5b, and sensors 1 and 2 in Fig. 2.5c. But if sensor 3 in Fig. 2.5c exchanges position with either sensor 1 or sensor 2, whether it maintains its phase or not, the signal across the output terminals will change. So for the purpose of determining how output signals can change with the exchange of sensor positions, basic topologies are a special case. So long as the relative phases are maintained, the positions of individual sensors within a basic topology makes no difference to the amplitude or phase or tone of the output of a basic topology.

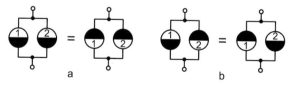

Fig. 2.6 The tonal equivalence of inverted phases: (**a**) in-phase & (**b**) out-of-phase

If a sensor circuit is flipped end-for-end across its terminals, as shown in Fig. 2.6a, b, it is extremely likely that the human ear can tell no difference in the tone. Why? Take a single sine wave signal, $\sin(2\pi ft)$, where f is frequency in cycles per second or Hertz (Hz), and t is time in seconds. The inverted signal, $-\sin(2\pi ft)$, is shifted in time by only a half cycle of the signal, as shown in Eq. 2.1.

$$-\sin\left(2\pi ft\right) = \sin\left(2\pi ft + \pi\right) = \sin\left(2\pi f\left(t + \frac{1}{2f}\right)\right) \qquad (2.1)$$

For middle C, or 440 Hz, that is a time difference of 1/880 second (s) or about 1.14 ms. Will the person who can show that the human ear has the equivalent of an atomic clock, and can hear the difference after the signal has been on for some time, please stand up and offer repeatable scientific proof? What about much more complex signals, with a number of tones, like the sound a guitar chord? Same question. In the absence of such proof, this work will assume that without any additional signal to act as a phase reference, the human ear cannot tell the difference between the signals of the 2 circuits in 2.6a, or the difference in the signals of the 2 circuits in 2.6b. In Fig. 2.6, $J = 2$, and $2^{J-1} = 2$ different phases, as shown.

There's another way to look at it. In linear circuit theory, all the components, such as resistors, capacitors, and inductors, have a linear response to time-varying, or AC, voltage signals. In Fig. 2.5c, say that the output at the terminals is $V_o(t)$, and the voltage signals generated by the sensors are $V1(t)$, $V2(t)$, and $V3(t)$. Then there are scalars, often dependent upon the frequencies of the signals, $a1(f)$, $a2(f)$, and $a3(f)$, such that $V_o(t) = a1(f) * V1(t) + a2(f) * V2(t) + a3(f) * V3(t)$. This is true for any circuit with these three sensors, although the scalars a1, a2, and a3 will vary from circuit to circuit. Each signal voltage has its own proportional effect on the output, V_o, regardless of the other voltages. If V1 and V2 have no output, then $V_o(t) = a3(f) * V3(t)$.

Now suppose that we add multipliers s1, s2, and s2, with only two values, +1 and −1, such that $V_o = s1 * a1 * V1 + s2 * a2 * V2 + s3 * a3 * V3$. If the value of an s-scalar is +1, then the sensor in the circuit is in normal phase, or in-phase. If the value of the s-scalar is −1, then the sensor is reversed in the circuit, connected out-of-phase, or contra-phase. Further, let us suppose that each of the s-scalars represents a digitally controlled 2P2T switch which can reverse the sensor connections within the circuit. When the control line c1, c2, or c3 for the switching of s1, s2, or s3, respectively, is a binary "0," then $s = +1$. When c is a binary "1" then $s = -1$. Start with all zeros for the c-lines, $(c3,c2,c1) = (0,0,0)$, and count up in binary, i.e., $(c3,c2,c1) = (0,0,0), (0,0,1), (0,1,0), (0,1,1), (1,0,0), (1,0,1), (1,1,0)$ and finally $(1,1,1)$. What's the result?

Table 2.1 Binary sequential switching of 3 sensor terminals between normal and reversed

(c3,c2, c1)	V_o		$-V_o$	(c3,c2, c1)
(0,0,0)	$a1 * V1 + a2 * V2 + a3 * V3$	=	$-(-a1 * V1 - a2 * V2 - a3 * V3)$	(1,1,1)
(0,0,1)	$-a1 * V1 + a2 * V2 + a3 * V3$	=	$-(a1 * V1 - a2 * V2 - a3 * V3)$	(1,1,0)
(0,1,0)	$a1 * V1 - a2 * V2 + a3 * V3$	=	$-(-a1 * V1 - a2 * V2 + a3 * V3)$	(1,0,1)
(0,1,1)	$-a1 * V1 - a2 * V2 + a3 * V3$	=	$-(a1 * V1 + a2 * V2 - a3 * V3)$	(1,0,0)

Notice that in each row of Table 2.1, the value of (c3,c2,c1) on the left is the complement of the value of (c3,c2,c1) on the right (0s and 1s switched). And each value of Vo on the left is the negative of the value of Vo on the right. There are $J = 3$ sensors here, and for $2^J = 2^3 = 8$ values of (c3,c2,c1), and the inability of the human ear to tell the difference between opposite phases of the output voltage, Vo(t), there are only $2^{J-1} = 2^2 = 4$ phase combinations that count. It does not matter if the circuit has a basic or complex topology, this will still hold.

As a consequence, for a circuit of J number of unmatched single-coil pickups, there are only 2^{J-1} number of unique phase reversals that the human ear can detect. But later, when we get to humbucking circuits made of matched-coil pickups, this will become more complicated.

2.6   *K* Sensors Taken *J* at a Time

Here, we need to have a brief review of combinatorial math. Equation 2.2 shows how to calculate the number of ways you can choose K things J at a time, where $J \leq K$ are both integers. For example, if you have 5 sensors and pick 2 of them to connect in series, you can do this in 10 different ways. If you have 3 sensors, A, B, and C, you can connect two of them in parallel in 3 different ways as: A‖B, A‖C, and B‖C, where B‖A is the same as A‖B. This kind of calculation is basic to this discussion (Eq. 2.2).

$$\binom{K}{J} = \frac{K!}{(K-J)! * J!} = \frac{K * (K-1) * \cdots * (K-J+1)}{J * (J-1) * \cdots * 2 * 1}, \quad K \geq J$$

$$\text{Examples:} \quad \binom{5}{2} = \frac{5 * 4}{2 * 1} = 10, \quad \binom{3}{2} = \frac{3 * 2}{2 * 1} = 3$$

(2.2)

Table 2.2 Combinations of *K* things taken *J* at a time

K	J=	1	2	3	4	5	6
1		1					
2		2	1				
3		3	3	1			
4		4	6	4	1		
5		5	10	10	5	1	
6		6	15	20	15	6	1
7		7	21	35	35	21	7
8		8	28	56	70	56	28
9		9	36	84	126	126	84
10		10	45	120	210	252	210

2.7   Constructing Series-Parallel Circuits

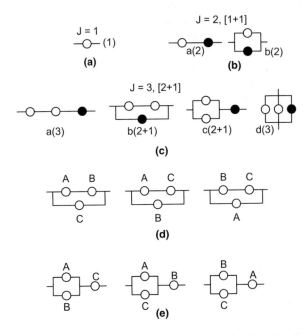

Fig. 2.7 Constructing circuits of size $J = 2$ and 3 from $J = 1$ and $J = 2$ circuits. The black discs are added sensors. Notations like $[2 + 1]$ refer to combinations of categories $J = 2$ and 1 to get category $J = 3$. Notations like $(2 + 1)$ refer to the kinds and numbers of basic topologies in the circuit constructed. (**a**) shows a single sensor. (**b**) shows two versions of category (2) circuits, a(2) and b(2). (**c**) Shows two versions of subcategory (3) circuits, a(3) and d(3), and two versions of subcategory $(2 + 1)$ circuits, b(2 + 1) and c(2 + 1). (**d, e**) Show how combinations of sensors in basic topologies are constructed, when different basic topologies are present. From Fig. 1a–e, Baker (2017a)

Figure 2.7a–e shows how series-parallel circuits of size $J = 1$ to 3 are systematically constructed. For this discussion, consider just the topologies of J things connected either in series or parallel, or some combination thereof, without changing the phases of individual sensors. This example also illustrates the definitions of some terms from Sect. 2.4, shown in **bold**.

For $J = 1$, there are no interconnections and the number of topologies is only $J = (1)$, where (1) represents a category of only 1 sensor connected between two terminals, as shown in Fig. 2.7a. For $J = 2$, there are only a series and a parallel connection, and the number of topologies is only 2, as shown in Fig. 2.7b. We construct this category (2) topology, labeled **a(2)** in Fig. 2.7b, merely by adding 1 sensor, designated by the filled circle, in series with 1 sensor, designated by the open circle. And by adding a category (1) sensor topology in parallel with another category (1) sensor topology, labeled **b(2)** in Fig. 2.7b. By this simple approach, adding equal and lower-category circuit topologies in series and parallel to existing circuit topologies, we construct all possible circuit topologies from previously existing topologies. Note that for category (2), two coils in series or two coils in parallel, the order of connection of the coils does not change the tone of the combination. This defines a **basic topology**. Hence, the use of (2) in the a(2) and b(2) notations.

For $J = 3$, we construct in Fig. 2.7c all possible **category** (3) topologies by adding the single **category** (1) topology in series and parallel with all possible **category** (2) topologies, indicated by the label $J = 3$, [2 + 1], where [2 + 1] indicates the use of **subcategories** (2) and (1) to construct topologies of 3 sensors. We see from inspection that this creates two **subcategories** of circuit topology, (3) and (2 + 1), which is the same as (3) and (1 + 2). The (2 + 1) in parentheses indicates that those topologies contains one **basic topology** of (2) and a **basic topology** of (1), connected together, as shown in Fig. 2.7c. We have a single sensor in parallel with a series pair, Fig. 2.7c-**b(2 + 1)** and a single sensor in series with a parallel pair, Fig. 2.7c-**c(2 + 1)**, for 2 **versions** of **subcategory** (2 + 1). Each **version** of (2 + 1) is constructed of a **basic topology** of category (2) connected to a **basic topology** of category (1). For subcategory (3), we have a **basic topology** of 3 sensors in series, Fig. 2.7c-**a(3)**, and a **basic topology** of 3 sensors in parallel, Fig. 2.7c-**d(3)**. This gives us 2 **versions** of (3).

Label the 3 sensors in Fig. 2.7c-**b(2 + 1)** as A, B, and C in Fig. 2.7d. By inspection, we can see that there are only 3 ways to connect these sensors together in this topology. Recall that order of connection does not matter in series or parallel **basic topologies**. The same is true for the topology in Fig. 2.7c-**c(2 + 1)**, as shown in Fig. 2.7e.

Equation 2.3a shows how this is done for each subcategory. For $J = 3$, subcategory (3), there is only 1 combination of 3 things taken 3 at a time. Changing the order of connection from A + B + C in series (using "+" to indicate series connections) to B + A + C cannot change the output. Changing the order of connection of A‖B‖C in parallel (using "‖" to indicate parallel connections) to C‖A‖B cannot change the output. For a subcategory like (2 + 1) with multiple basic topologies, the combination calculations must be split up and multiplied together. First the (2) part is calculated by taking 3 things 2 at a time, then 2 is subtracted from 3, leaving 1 thing taken 1 at a time. Equation 2.3b shows the combinations in each subcategory multiplied by the number of versions in each subcategory, adding up to $J_T = 8$ total unique topologies, comprising of Fig. 2.7c-**a(3)**, or A + B + C in series, Fig. 2.7c-**d(3)**, or A‖B‖C in parallel, plus the combinations in Fig. 2.7d & e. So J_T in this context is an important number in topologies of all sizes of J, including combinatorial placements of sensors within all the versions of topologies. When there are K sensors to be taken J at a time, that combination multiplies by J_T to calculate the total number of ways that K sensors can be combined in a set of topologies of size J or category (J), as shown in Eq. (2.3c), not counting phase changes.

$$3: \binom{3}{3} = 1$$

(2.3a)

$$2 + 1: \binom{3}{2} * \binom{3-2}{1} = \binom{3}{2} * \binom{1}{1} = 3 * 1 = 3$$

$$J_T = \binom{3}{3} * 2 + \binom{3}{2} * \binom{1}{1} * 2 = 1 * 2 + 3 * 2 = 8 \qquad (2.3b)$$

$$\text{Total number of possible connections} = \binom{K}{J} * J_T \qquad (2.3c)$$

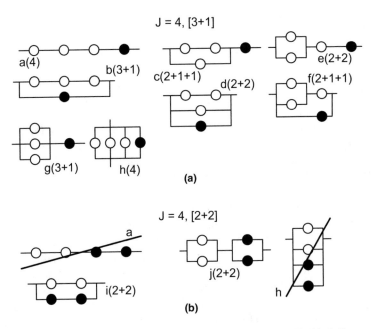

Fig. 2.8 Constructing $J = 4$ sensor circuits from $J = 1, 2,$ and 3 circuits. The black discs are added sensors, such as adding category $J = 1$ to category $J = 3$ (**a**), and $J = 2$ to $J = 2$(**b**). The brackets [] show what categories are added together, and the parentheses () show the resulting collections of basic topologies in the $J = 4$ categories constructed. The slashes show topologies duplicate to those already constructed, by visual inspection. It should be obvious that adding categories in the reverse order, $[1 + 3]$, would produce the same results as (**a**), with a lot more black discs. From Fig. 2a, b, Baker (2017a)

Figure 2.8a, labeled $J = 4, [3 + 1]$, shows how the $J = 1$ category topology is combined with $J = 3$ category topologies, to obtain the $(3 + 1)$ subcategory for $J = 4$ category topologies. For example, Fig. 2.8a-**a(4) and b(3 + 1)** shows a single sensor connected in series and parallel with the topology in Fig. 2.7c-**a(3)** to obtain 2 new $J = 4$ topologies. As is done with the remaining Fig. 2.7c topologies to obtain a total of 2(4), 2(3 + 1), 2(2 + 2) and 2(2 + 1 + 1) subcategory versions. Figure 2.8b shows how both category (2) topologies are combined to produce 2 new (2 + 2) versions and 2 topologies already constructed in Fig. 2.8a, which duplicate Fig. 2.8a-**a(4)** and **h(4)**. Altogether, there are 4(2 + 2) subcategory versions in the $J = 4$ topologies, for a total of 10 versions of $J = 4$ topologies.

We find that in doing so, the topologies for subcategory (2 + 1 + 1) are also constructed. For subcategory (2 + 1 + 1), two single sensor basic topologies are connected to a basic topology serial or parallel pair of sensors, in such a way that the order of choice of the single coils matters to the tone, which we can see by inspection. Equation 2.4a shows the number of tonal combinations of $J = 4$ sensors for each version of topology in a subcategory in Fig. 2.8, (4), (3 + 1), (2 + 2) and (2 + 1 + 1). Note how the lower numbers in each of the bracketed combinatorial expressions match the numbers between the parentheses in the subcategory labels. Equation 2.4b shows the number of combinations of sensors times the number of versions of topology in each subcategory to obtain the total number, $J_T = 58$ unique topologies, from 4 subcategories.

$$4: \binom{4}{4} = 1$$

$$3 + 1: \binom{4}{3} * \binom{4-3}{1} = \binom{4}{3} * \binom{1}{1} = 4 * 1 = 4$$

$$2 + 2: \binom{4}{2} * \binom{4-2}{2} = \binom{4}{2} * \binom{2}{2} = 6 * 1 = 6$$

$$2 + 1 + 1: \binom{4}{2} * \binom{4-2}{1} * \binom{4-3}{1} = \binom{4}{2} * \binom{2}{1} * \binom{1}{1} = 6 * 2 * 1 = 12$$

$$(2.4a)$$

$$J_T = \binom{4}{4} * 2 + \binom{4}{3} * \binom{1}{1} * 2 + \binom{4}{2} * \binom{2}{2} * 4 + \binom{4}{2} * \binom{2}{1} * \binom{1}{1} * 2$$

$$J_T = 1 * 2 + 4 * 2 + 6 * 4 + 12 * 2 = 58$$

$$(2.4b)$$

We might have considered combinatorial placements of sensors in each version of a subcategory as a separate topic, to be discussed after generating all the versions of subcategories for all Js. But this way the overall calculation for K sensors, taken J at a time, without phase changes, is a simpler and more easily understood equation.

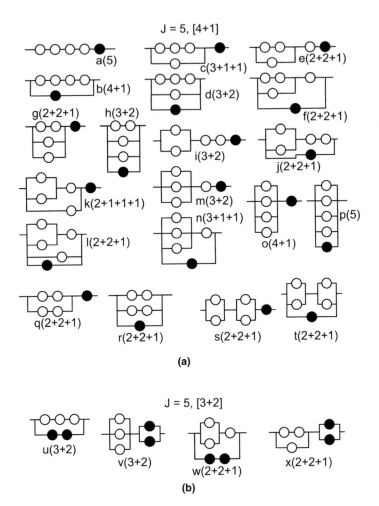

Fig. 2.9 Construction of category $J = 5$ circuits from categories $[4 + 1]$ (**a**) and $[3 + 2]$ (**b**), producing 2 versions of subcategory (5), 2 of $(4 + 1)$, 6 of $(3 + 2)$, 2 of $(3 + 1 + 1)$, 11 of $(2 + 2 + 1)$, and 1 of $(2 + 1 + 1 + 1)$, for a total of 24. Slashed-out duplicates are not shown. From Fig. 3a, b, Baker (2017a)

Figure 2.9 shows the constructions of topologies for $J = 5$, not showing any of the duplicate circuits that were generated, identified by inspection, and discarded. This is left as an exercise for the reader. Figure 2.9a, labeled $J = 5$, $[4 + 1]$, shows those constructed from topological categories (4) and (1). Note that there are 20 new topologies, as one might expect from adding (1) in series and parallel with the 10 topologies of $J = 4$. This produces 2 versions of subcategory (5), 2 of $(4 + 1)$, 4 of $(3 + 2)$, 2 of $(3 + 1 + 1)$, 9 of $(2 + 2 + 1)$, and 1 of $(2 + 1 + 1 + 1)$.

Figure 2.9b, labeled $J = 5, [3 + 2]$, shows those constructed from topological categories (3) and (2), leaving out all those previously constructed. Altogether, this produces 2 versions of subcategory (5), 2 versions of $(4 + 1)$, 6 versions of $(3 + 2)$, 2 versions of $(3 + 1 + 1)$, 10 versions of $(2 + 2 + 1)$, and 2 versions of $(2 + 1 + 1 + 1)$, for 24 versions of $J = 5$ subcategory topologies. Equation 2.5a shows numbers of combinations of $J = 5$ sensors for each of the subcategories, and Eq. (2.5b) shows their products times the number of versions in each subcategory, for a total of $J_T = 502$ unique topologies, from 6 subcategories.

$$5: \binom{5}{5} = 1$$

$$4+1: \binom{5}{4} * \binom{5-4}{1} = \binom{5}{4} * \binom{1}{1} = 5*1 = 5$$

$$3+2: \binom{5}{3} * \binom{5-3}{2} = \binom{5}{3} * \binom{2}{2} = 10*1 = 10$$

$$3+1+1: \binom{5}{3} * \binom{5-3}{1} * \binom{5-4}{1} = \binom{5}{3} * \binom{2}{1} * \binom{1}{1} = 10*2*1 = 20$$

$$2+2+1: \binom{5}{2} * \binom{5-2}{2} * \binom{5-4}{1} = \binom{5}{2} * \binom{3}{2} * \binom{1}{1} = 10*3*1 = 30$$

$$2+1+1+1: \binom{5}{2} * \binom{5-2}{1} * \binom{5-3}{1} * \binom{5-4}{1} = \binom{5}{2} * \binom{3}{1} * \binom{2}{1} * \binom{1}{1} = 10*3*2*1 = 60$$

$$(2.5a)$$

$$J_T = \binom{5}{5} * 2 + \binom{5}{4} * \binom{1}{1} * 2 + \binom{5}{3} * \binom{2}{2} * 6 + \binom{5}{3} * \binom{2}{1} * \binom{1}{1} * 2$$
$$+ \binom{5}{2} * \binom{3}{2} * \binom{1}{1} * 11 + \binom{5}{2} * \binom{3}{1} * \binom{2}{1} * \binom{1}{1} * 1$$
$$J_T = 1*2 + 5*2 + 10*6 + 20*2 + 30*11 + 60*1 = 502$$

$$(2.5b)$$

Without further mathematical demonstration or proof, one may offer the conjecture that in constructing topologies, i.e., for J number of sensors, using topological categories for (J) and smaller, that one only needs to do the constructions from pairs of smaller categories, i.e., $(J - 1)$ and (1), then $(J - 2)$ and (2), down to $(J - n)$ and (n), where $(J - n)$ is an integer greater than or equal to $J/2$. That from these combinations, all the other subcategories with 3 or more basic topologies are also created, i.e., $((J - 2) + 1 + 1)$, $((J - 3) + 2 + 1)$, $((J - 3) + 1 + 1 + 1)$, and others.

The 5 pages of laboratory notebook constructions for $J = 6$ circuits can be found at: https://www.researchgate.net/publication/323390784_On_the_Topologies_of_ Guitar_Pickup_Circuits

The original text of that document had a math error in the equation relating to Eq. (2.5b), incorrectly finding that $J_T = 532$. Math errors are always possible, and valid corrections are appreciated.

The constructions for $J = 6$, from [5 + 1], [4 + 2], and [3 + 3] produced circuits labeled from: a through z, aa through zz, and aaa through ttt. Unless some error was made, it produced the following versions (Table 2.3):

Table 2.3 Results of constructions for $J = 6$, from $J = (1)$, (2), (3), (4), and (5)

Subcategories	Members	#Members
(6)	a,p	2
(5 + 1)	b,o	2
(4 + 2)	d,m,q,zz,aaa	5
(3 + 3)	c,n	2
(4 + 1 + 1)	h,i,nnn,qqq	4
(3 + 2 + 1)	e,q,j,l,r,x,y,ee,ff,oo,pp,qq,rr,ww,xx,yy,ooo,ppp	18
(3 + 1 + 1 + 1)	f,k	2
(2 + 2 + 2)	t,cc,gg,jj,kk,nn,bbb,fff,ggg,hhh,iii,jjj,kkk,lll,mmm	15
(2 + 2 + 1 + 1)	s,u,w,z,bb,dd,hh,ii,ll,mm,ss,tt,uu,vv,ccc,ddd,eee,rrr,sss,ttt	20
(2 + 1 + 1 + 1 + 1)	v,aa	2
	Versions, Jv =	72

Equation 2.6a shows numbers of combinations of $J = 6$ sensors for each of the subcategories, and Eq. (2.6b) shows their products times the number of versions in each subcategory, for a total of $J_T = 7219$ unique topologies, from 10 subcategories.

$$6: \binom{6}{6} = 1$$

$$5 + 1: \binom{6}{5} * \binom{1}{1} = 6 * 1 = 6$$

$$4 + 2: \binom{6}{4} * \binom{2}{2} = 15 * 1 = 15$$

$$4 + 1 + 1: \binom{6}{4} * \binom{2}{1} * \binom{1}{1} = 15 * 2 * 1 = 30$$

$$3 + 3: \binom{6}{3} * \binom{3}{3} = 20 * 1 = 20$$

$$3 + 2 + 1: \binom{6}{3} * \binom{3}{2} * \binom{1}{1} = 20 * 3 * 1 = 60$$

$$3 + 1 + 1 + 1: \binom{6}{3} * \binom{3}{1} * \binom{2}{1} * \binom{1}{1} = 20 * 3 * 2 * 1 = 120$$

$$2 + 2 + 2: \binom{6}{2} * \binom{4}{2} * \binom{2}{2} = 15 * 6 * 1 = 90$$

$$2 + 2 + 1 + 1: \binom{6}{2} * \binom{4}{2} * \binom{2}{1} * \binom{1}{1} = 15 * 6 * 2 * 1 = 180$$

$$2 + 1 + 1 + 1 + 1: \binom{6}{2} * \binom{4}{1} * \binom{3}{1} * \binom{2}{1} * \binom{1}{1} = 15 * 4 * 3 * 2 * 1 = 360$$

$$(2.6a)$$

$$\begin{aligned} J_T = {} & 1 * 2 + 6 * 2 + 15 * 5 + 30 * 2 + 20 * 4 + 60 * 18 \\ & + 120 * 2 + 90 * 15 + 180 * 20 + 360 * 2 = 7219 \end{aligned}$$

$$(2.6b)$$

For $J = 7$, no topologies have been constructed here, but it is reasonable to suppose that they may be constructed from combining category (6) topologies with category (1), (5) with (2), and (4) with (3), likely producing about 14 subcategories: (7), (6 + 1), (5 + 2), (4 + 3), (5 + 1 + 1), (4 + 2 + 1), (4 + 1 + 1 + 1), (3 + 3 + 1), (3 + 2 + 2), (3 + 2 + 1 + 1), (3 + 1 + 1 + 1 + 1), (2 + 2 + 2 + 1), (2 + 2 + 1 + 1 + 1), and (2 + 1 + 1 + 1 + 1 + 1). Some extrapolations have been made using curve-fitting, which is not very precise. Several fits suggest that for $J = 7$: $267 \leq J_V \leq 407$; $42{,}300 \leq J_T \leq 207{,}200$; and $10.4(10^6) \leq J_T * N_{SGN} \leq 11.3(10^6)$. In any case, it's a lot.

Now that the method of construction has been established, it is no doubt possible to continue and expand it to any number of J greater than 6 by computer program. Possibly using SPICE notation in the constructions would make it easier to find and delete the inevitably constructed duplicate circuits. It might be as simple as constructing and representing the circuits as text strings with nested parentheses,

"+" for series connections and "‖" for parallel connections, then having a computer program analyze the strings for basic topologies and duplicate circuits. But this author must leave it for others to accomplish; certain medications have made that kind of work too difficult.

2.8 Collecting Results

Recall from Sect. 2.5 that the number of possible terminal reversals of individual sensors is 2^{J-1}. Call this number $N_{SGN} = 2^{J-1}$. For single-sensor circuits of size J, with a total number of topology versions, J_T, including the placement of individual sensors in different positions of each topology, J_T must be multiplied by N_{SGN} to include the number of ways that the circuit output phase and tone can be changed by reversing the terminals in individual sensors. Then for K sensors, $K > J$, that number must be multiplied by K things taken J at a time, as shown by K_{JT} in Eq. 2.6a. Equation 2.3b shows the total number of unique circuits of size $J \leq K$, K_T. Table 2.4 shows the collected results for K number of sensors from 1 to 10 and category J circuits from 1 to 6.

$$K_{JT} = \binom{K}{J} * N_{SGN} * J_T = \binom{K}{J} * 2^{J-1} * J_T \qquad (2.7a)$$

$$K_T = \sum_{J=1}^{K} K_{JT} = \sum_{J=1}^{K} \binom{K}{J} * 2^{J-1} * J_T \qquad (2.7b)$$

Table 2.4 The collected results for K number of sensors from 1 to 10 and category J circuits from 1 to 6, where J_V is the number of versions, J_T is the total number of combinations of sensors and versions, and K_T is the total number of circuits of size $J = 1\text{--}6$, for $K \geq J$

J	1	2	3	4	5	6	
J_V	1	2	4	10	24	72	
J_T	1	2	8	58	502	7219	
N_{SGN}	1	2	4	8	16	32	
$J_T * N_{SGN}$	1	4	32	464	8032	231,008	
K							K_T
1	1						1
2	2	4					6
3	3	12	32				47
4	4	24	128	464			620
5	5	40	320	2320	8032		10,717
6	6	60	640	6960	48,192	231,008	286,866
7	7	84	1120	16,240	168,672	1,617,056	1,803,179
8	8	112	1792	32,480	449,792	6,468,224	6,952,408
9	9	144	2688	58,464	1,012,032	19,404,672	20,478,009
10	10	180	3840	97,440	2,024,064	48,511,680	50,637,214

It is unlikely that a standard electric guitar with a 64.8 cm (25.5″) baselength could hold more than 8 or 9 single-coil pickups between the neck and bridge. And by some experiments, the different tones tend to bunch at the warm end, some so close together as not to count as distinct tones. So these numbers can only be counted as "potentially unique tones." This is likely much more so for higher numbers of pickups closer together, where their fields are likely to entangle and possibly interfere. So the numbers of K up to 10 would likely make sense only for pickups placed along the length of piano strings.

The distance between poles of a humbucker is about 2 cm (0.8″). For a guitar of standard baselength, this amounts to about 32nd harmonic of the string fretted at the nut, and about the 16th harmonic of a string fretted at 12. Considering how fast harmonics die off in amplitude, it remains to be seen, proven only be experiment and measurement, whether or not pickups so close together can generate very many unique tones.

Also later developments here will show that only a small percentage or fraction of a percent of these huge numbers of circuits can be humbucking, and then only with pickups matched to have equal responses to external hum. At the time of this writing, no pickups other than humbuckers are made or offered for sale with that specification. The time of matched single-coil pickups has yet to come.

References

Patents

Damm, W. (2002). Single-coil electric guitar pickup with humbucking-sized housing, US Patent 6,372,976 B2, 16 Apr 2002. Retrieved from https://patents.google.com/patent/US6372976B2/

Fender, C. L. (1966). Electric guitar incorporating improved electromagnetic pickup assembly, and improved circuit means, US Patent 3,290,424, 6 Dec 1966. Retrieved from https://patents.google.com/patent/US3290424A/

Fender, C. L. & Kaufmann, C. O. (1948). Pickup unit for stringed instruments, US Patent 2,455,575, 7 Dec 1948. Retrieved from https://patents.google.com/patent/US2455575A/

Furst. W. & Boxer, M. (2001). Sound pickup switching apparatus for a string instrument having a plurality of sound pickups, US Patent 6,316,713 B1, 12 Nov 2001. Retrieved from https://patents.google.com/patent/US6316713B1/

Hamilton, J. W. (2011). Three pickup guitar switching system with two options, US Patent 7,999,171 B1, 16 Aug 2011. Retrieved from https://patents.google.com/patent/US7999171B1/

Knapp, L. J. (1994). Electronic guitar equipped with asymmetrical humbucking electromagnetic pickup, US Patent 5,292,998, 8 Mar 1994. Retrieved from https://patents.google.com/patent/US5292998A/

Lace, M. A. (1995a). Electromagnetic musical pickup using main and auxiliary permanent magnets, US Patent 5,389,731, 14 Feb 1995. Retrieved from https://patents.google.com/patent/US5389731A/

Lace, M. A. (1995b). Electromagnetic musical pickups with central permanent magnets, US Patent 5,408,043, 18 Apr 1995. Retrieved from https://patents.google.com/patent/US5408043A/

Miessner, B. F. (1933). Method and apparatus for the production of music, US Patent 1,915,858, 27 June 1933. Retrieved from https://patents.google.com/patent/US1915858A/

Morrison, G. E. (1951). Magnetic pickup unit for guitars, US Patent 2,557,754, 19 June 1951. Retrieved from https://patents.google.com/patent/US2557754A/

Olvera, J. C. & Olvera, G. A. (2004). Electric guitar circuit control and switching module, US Patent 6,781,050 B2, 24 Aug 2004. Retrieved from https://patents.google.com/patent/US6781050B2/

Peavey, H. D. (1981). Selector switch, US Patent 4,305,320, 15 Dec 1981. Retrieved from https://patents.google.com/patent/US4305320A/

Riboloff, J. T. (1994). Guitar pickup system for selecting from multiple tonalities, US Patent 5,311,806, 17 May 1994. Retrieved from https://patents.google.com/patent/US5311806A/

Saunders, J. H. (1989). Control system with memory for electric guitars, US Patent 4,817,486, 4 Apr 1989. Retrieved from https://patents.google.com/patent/US4817486A/

Schaller, H. F. K. (1985). Magnetic pickup for stringed instruments, US Patent 4,535,668, 20 Aug 1985. Retrieved from https://patents.google.com/patent/US4535668A/

Simon, J. C. (1979). System for selection and phase control of humbucking coils in guitar pickups, US Patent 4,175,462, 27 Nov 1979. Retrieved from https://patents.google.com/patent/US4175462A/

Starr, H. W. (1987). Electric guitar pickup switching system, US Patent 4,711,149, 8 Dec 1987. Retrieved from https://patents.google.com/patent/US4711149A/

Stich, W. L. (1975). Electrical pickup for a stringed musical instrument, US Patent 3,916,751, 4 Nov 1975. Retrieved from https://patents.google.com/patent/US3916751A/

Thompson, P. G. (1998). Switching apparatus for electric guitar pickups, US Patent 5,763,808, 9 June 1998. Retrieved from https://patents.google.com/patent/US5763808A/

Wnorowski, T. F. (2006). Method for switching electric guitar pickups, US Patent 6,998,529 B2, 14 Feb 2006. Retrieved from https://patents.google.com/patent/US6998529B2/

Wolstein, R. J. (1992). Guitar pickup and switching apparatus, US Patent 5,136,919, 11 Aug 1992. Retrieved from https://patents.google.com/patent/US5136919A/

Patent Applications

Baker, D. L. (2017a). Humbucking switching arrangements and methods for stringed instrument pickups, US Patent Application 15/616,396, filed 7 June 2017, published as US-2018-0357993-A1, 13 Dec 2018, granted as Patent US10,217,450, 26 Feb 2019. Retrieved from https://www.researchgate.net/publication/335727402_Humbucking_switching_arrangements_and_methods_for_stringed_instrument_pickups_-_NPPA_15616396

Jacob, B. L. (2009). Programmable switch for configuring circuit topologies, US Patent Application 2009/0308233 A1, 17 Dec 2009. Retrieved from https://patents.google.com/patent/US20090308233A1/

Krozack, E., et al. (2005). Multi-mode multi-coil pickup and pickup system for stringed musical instruments, US Patent Application 2005/0150364A1, 14 July 2005. Retrieved from https://patents.google.com/patent/US20050150364A1/

Chapter 3
Series-Parallel Circuit Topologies of Humbucking Pickups

3.1 Some Prior Art

Dual-coil humbucking pickups generally have coils of equal matched turns, as demonstrated in the patents of Lesti (1936, US2026841), Lover (1959, US2896491), Fender (1961, US2976755), and Blucher (1985, US4501185). The 1959 Lover patent, assigned to Gibson Inc., looks and acts very much like modern humbuckers. At least one patent describes a dual-coil humbucker with one coil and poles adjacent the strings, and the other vertically in line and below (Anderson, 1992, US5168117), sometimes called "stacked coils." One hears of single-coil pickups with an added sensor coil of a different numbers of turns which detect a hum signal to be subtracted from the output so as to cancel hum.

Humbuckers with two matched coils can have those coils connected in either series or parallel. Individual humbuckers commonly have either 4 wires, 2 for each coil, or 2 wires, with the coils connected in series for maximum voltage output, often with a shield wire connected to the metal parts of the humbucker and the pickup cable shield. Guitars with two humbuckers commonly have a 3-way switch, which offers for output the bridge humbucker, the neck humbucker, and the two connected in parallel in the middle switch position. Some guitars combine two humbuckers, one at the neck and one at the bridge, with a single-coil pickup mounted in between them. Some use as many as 3 humbuckers.

Electric bass guitars are another matter, often containing only two single-coil pickups, or a single-coil pickup with two single-coil pickups at one position, split between pairs of strings, which can be nominally humbucking. Those are not addressed here, though the same principles can apply for pickups and circuits made as described here.

© Springer Nature Switzerland AG 2020
D. L. Baker, *Sensor Circuits and Switching for Stringed Instruments*,
https://doi.org/10.1007/978-3-030-23124-8_3

3.2   Adapting Single-Coil Circuits to Dual-Coil Humbuckers

If one considers using only humbucking electromagnetic pickups as single sensors, without combining single coils from different humbuckers, it is possible to use the same topologies developed in Chap. 2, replacing each sensor or single-coil pickup (Fig. 3.1a) with a dual-coil humbucker, as in Fig. 3.1, where the humbucker coils may be connected in parallel (Fig. 3.1-b(2)) or series (Fig. 3.1-c(2)), but the equations have to be modified. The total number of versions are $JJ_T = J_T$, since the single-coil pickups are replaced by dual-coil humbuckers in the same topologies and subcategories. The individual coils in each humbucker can be connected either in series or parallel, giving 2 choices of sub-combination for each humbucker, as expressed in JJ_{SP} in Math 3.1a. Math 2.6a-b then becomes Math 3.1b-c (Table 3.1).

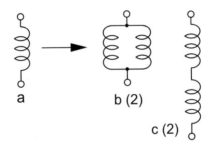

a   b (2)

c (2)

Fig. 3.1 Exchanging parallel- and series-connected HB coils for a single coil to extend the single-coil topologies to dual-coil humbuckers. From Fig. 4, Baker (2017a)

$$JJ_{SP} = 2^{JJ} \tag{3.1a}$$

$$KK_{JJT} = \binom{KK}{JJ} * N_{SGN} * JJ_{SP} * JJ_T, \quad KK \geq JJ$$

$$KK_{JJT} = \binom{KK}{JJ} * 2^{JJ-1} * 2^{JJ} * JJ_T, \quad KK \geq JJ \tag{3.1b}$$

$$KK_T = \sum_{JJ=1}^{KK} KK_{JJT} = \sum_{JJ=1}^{KK} \binom{KK}{JJ} * 2^{JJ-1} * 2^{JJ} * JJ_T \tag{3.1c}$$

Table 3.1 Total number of possibly unique tones from KK humbuckers taken JJ at a time, with sums KK_T across all possible $JJ \le KK$

JJ	1	2	3	4	5	
N_{SGN}	1	2	4	8	16	
JJ_{SP}	2	4	8	16	32	
$JJ_T = J_T$	1	2	8	58	502	
$N_{SGN} * JJ_{SP} * JJ_T$	2	16	256	7424	257,024	
KK	KK_{JJT}					KK_T
1	2					2
2	4	16				20
3	6	48	256			310
4	8	96	1024	7424		8552
5	10	160	2560	37,120	257,024	296,874

From Math 14, Baker (2017a)

3.3 A Dual-Humbucker Experiment

Fig. 3.2 Modified electric guitar with two generic Hofner-style mini-humbuckers, mounted in Brazilian Cherry plates, with 24-way switching circuit made from two 4P2T toggle switches (top right of bridge HB mounting plate) and a 4P6T rotary switch, producing 20 different coil circuit topologies, and 16–17 unique tones. The toggle switch below the rotary switch had no function, and the brown knob below that controlled a 500k volume pot

3.3.1 Basic Measurements of the Pickups

Figure 3.2 shows the modified electric guitar used in the experiment. As noted in the caption, it had 2 generic mini-humbuckers. The internal coils connected in series had a resistance of about 6.62 kΩ. Which means that each coil had a resistance of about 3.31 kΩ. As shown below, the coil inductance was a bit equivocal. Each pickup had a single bar magnet connected to the poles, providing a South pole for rectangular flat poles flush with the pickup cover, and a North pole for the round-head adjustable screw poles rising above the pickup cover. In Fig. 3.2, the screw poles are on the neck side of the pickup, and the flat poles on the bridge side (Fig. 3.3).

Fig. 3.3 Top and bottom view of the generic Hofner-type mini-humbuckers, mounted on Brazilian Cherry plates, with 4-wire shielded cable

The pickups came with the internal coils connected in series and in phase. The output was a coaxial cable, to which the ground was connected to the bottom end of the S-up coil and the metal frame of the pickup, and the center conductor to the top end of the N-up coil. The HB was disassembled and rewired to a 4-wire shielded output, using Alpha 1214C cable, which turned out to be a little stiff for easy routing under a pick guard. The original high side of the N-up (flat pole) coil was taken to be the + side; the interconnection to the S-up (screw pole) coil was taken to be the + side of that coil. The color code connections to the Alpha cable are: Red = N+, Green = N−, White = S+, Black = S−, and shield = pickup metal frame.

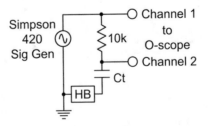

Fig. 3.4 Resonant notch filter measurement circuit of humbucker (HB) inductance, using oscilloscope Chan1-Chan2, which drops in value when the signal generator frequency matches the resonance of capacitor Ct with the HB inductance, according to Eq. (3.2)

$$2\pi f = \frac{1}{\sqrt{LC}} \tag{3.2}$$

where f = frequency (Hz), L = inductance (H), and C = capacitance (F).

For low frequencies, the impedance of Ct in series with the humbucker is high because of Ct. For high frequencies, it is high because of the inductance of the humbucker. It is lowest at the resonant frequency, making the difference signal across the 10k resistor the highest, as measured on the scope display Channel 1 minus Channel 1 (Table 3.2).

Table 3.2 Calculations of humbucker inductance in Fig. 3.4 from resonant frequency measurements and Eq. (3.2) for series- and parallel-connected humbucker coils

Ct (nF)	Series f (Hz)	Parallel f (Hz)	Series L (H)	Parallel L (H)
1	3490	7140	2.08	0.50
1.5	3125		1.73	
4.7		3160		0.54
10	920	2010	2.99	0.63
15	764	1676	2.89	0.60
22	606	1319	3.14	0.66
33	512	1105	2.93	0.63
47	390	860	3.54	0.73
68	322	656	3.59	0.87
100	250	532	4.05	0.89
220	200	380	2.88	0.80
		Mean =	2.98	0.68
		Std dev =	0.69	0.13

If the coils of the humbucker were independent of each other, the inductances of the series and parallel connections would have a ratio of 4 to 1. But the mean values of the measurements have a ratio of 2.98 to 0.68 H, which is a 4.38 to 1. The measurements their standard deviations show a lot of variation. Part of this could be ascribed to the 10% tolerance of the values of the ceramic capacitors used.

But some may be due to the electromagnetic coupling between the coils, due to a common magnetic circuit. In one experiment with another type of humbucker, a signal applied to one coil produced a much smaller signal in the other, making it a weak transformer. But this author has not done such calculations in decades and must leave it to others to investigate. This particular measurement also fails to take into account the winding capacitance of the coils.

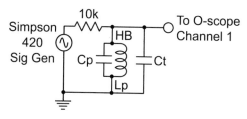

Fig. 3.5 Resonant peak filter for parallel-connected dual-coil humbucker, where Cp = parallel coil capacitance (nF), Lp = parallel coil inductance (H), and Ct = test capacitance (nF)

Figure 3.5 shows a better setup for that. The unknown coil capacitance is $Cc = Cp/2$. The unknown coil inductance is $Lc = 2 * Lp$. In this case, at resonance the impedance of the humbucker, HB, is at a maximum, and shows as a peak in output on oscilloscope Channel 1, which shares the same ground as this circuit.

$$Lp = \frac{1}{(Cp + Ct) * (2\pi f)^2} \tag{3.3}$$

where Lp = HB inductance (H), Cp = HB capacitance (F), Ct = test cap (F), f = resonant frequency (Hz).

Table 3.3 Results from a spreadsheet, where Cp (0.225 nF) is estimated for the calculation of Eq. (3.3), then optimized by minimizing the sum of the standard deviation (0.144) of all the calculated Lp plus the square of the difference between the mean Lp (0.583 H) and the Lp (0.601) for Ct = 0

Cp =	2.25E−10	0.225		Equiv coil Cc =	0.112
				Equiv coil Lc =	1.167
Ct (nF)	Ct (F)	∥ f (Hz)	Lp calc (H)		
0	0	13,699	0.601	0.00030477	
0.68	6.8E−10	8511	0.387		
1	1E−09	6704	0.460		
1.5	1.5E−09	5556	0.476		
2.2	2.2E−09	5128	0.397		
3.3	3.3E−09	3659	0.537		
4.7	4.7E−09	3141	0.521		
6.8	6.8E−09	2448	0.602		
10	1E−08	2051	0.589		
15	1.5E−08	1604	0.647		
22	2.2E−08	1225	0.760		
33	3.3E−08	1020	0.733		
100	1E−07	537	0.876		
		Mean =	0.583		
		Std dev =	0.144	0.1446524	

Table 3.3 shows the spreadsheet results for calculating the mean of Lp calculated for all of the test values of Ct, using Eq. (3.3). First Cp is set to an estimate, such as 0.1 nF. Then the mean and standard deviation are taken from the calculated values of Lp. The value of 0.00030477 is the square of (0.601 − 0.583). This is added to the standard deviation (0.1446524), and the spreadsheet Solver minimizes this number by adjusting Cp. Then the individual coil capacitance, Cc, is set to half Cp, because the capacitances of the coils are in parallel, and the individual coil inductance, Lc, is set to twice the mean Lp, because the coil inductances are in parallel. And the standard deviation of calculated inductances tells us how good those numbers are.

The results here can only be taken as representative of this experiment, and may not be freely transferred to any other. This seems somewhat imprecise and unsatisfactory. And so it is. But it helps to find out what is possible, especially with limited resources, like a voltmeter, a signal generator, an oscilloscope, and a shareware FFT program that only produces real-valued amplitude spectra, leaving out the phase information. This 24-way switching system can produce 20 different circuits from 2 humbuckers, but cannot be wired to put the tones in order from bright to warm. The next experiment after that led to a new way of switching pickups that can produce all-humbucking tones, which can be ordered from bright to warm.

This is the value of research, learning from failures and partial failures to design the next experiment to see what can be done. Considering that these measurements used only a signal generator and an oscilloscope, with 10% precision test capacitors, others can no doubt come up with better methods and numbers.

3.3.2 The 24-Way Switching Circuit

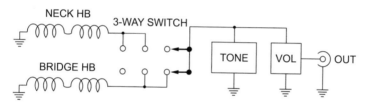

Fig. 3.6 Symbolic electrical representation of the common 3-way switching circuit for a dual-humbucker guitar

For reference, Fig. 3.6 shows a symbolic representation of a common 3-way switching circuit for a dual-humbucker guitar. An actual 3-way switch is built somewhat differently than the 2P3T switch shown here, as a symbolic convenience. It connects the bridge HB to the output, followed by the bridge and neck HBs in parallel, then the neck HB alone.

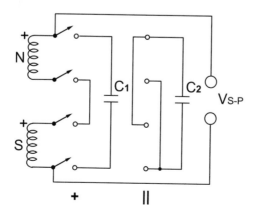

Fig. 3.7 4P2T Series-parallel switch for one humbucker, with peaking capacitors $C1$ and $C2$, and series-parallel output, V_{S-P}. "+" indicates the series connection and "||" indicates the parallel connection

Figure 3.7 shows the 4P2T series-parallel switch circuit used on each humbucker. This experiment was run twice, with no capacitors, $C1$ and $C2$, and with $C1 = 3.3$ nF and $C2 = 10$ nF. The same circuit could be accomplished with a 3P2T switch, with all the lower pole and throws connected directly to the lower terminal of V_{S-P}. But the four poles made convenient terminals for soldering the wires.

The other part might come from the fact that, due to having a common magnetic circuit, the two coils are weakly coupled. This author recalls having fed a signal into one coil of a humbucker, not necessarily this kind, and gotten a weak signal out of the other, making a dual-coil humbucking a weak transformer. But this author will leave the analysis of such effects to others, not having done it himself in decades.

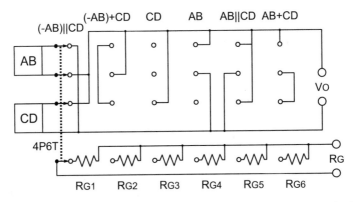

Fig. 3.8 Use of a 4P6T rotary switch to combine 2 sensors/pickups into 6 unique circuits, where "−" indicates reversed terminals/phase, "+" indicates a series connection, "‖" indicates a parallel connection, with one pole and set of throws being used to switch other components, such as resistors, R_{Gi}, for gain adjustments. From Fig. 18, Baker (2017a)

The 2-terminal outputs, V_{S-P}, shown as AB and CD in Fig. 3.8, each indicating the two coils in one humbucker, connects to the next switch, a 4P6T switch to produce 2 parallel, 2 series and 2 single pickup circuits from them. Each of the switch circuits in Fig. 3.7, one for each HB, have 2 configurations, making the total number equal to $2 * 2 * 6 = 24$ switch configurations. But for the two configurations in Fig. 3.8 where only one pickup is used, the switching in Fig. 3.7 for the other pickup has no effect. So the number of unique circuits produced by this switching system is $24 - 4 = 20$ circuits, all humbucking. Compare this to not just the 3-way switch in Fig. 3.6, but to Fender's (1966, US3290424) switching system for 4 coils (Figs. 2.1 and 2.2 in Chap. 2) with 81 switch configurations, which produced only 9 unique humbucking circuits, plus their reversed-terminal opposite phases, of which the human ear cannot be expected to hear the difference.

3.3.3   Practical Results

How can an engineer measure the difference between bright and warm or dark tones coming from a particular pickup circuit? Will it make any sense to musicians? Can it be used to order tones in a pickup switching system? A fair amount of work still needs to be done on those questions. French (2009)[1] mentions A-weighting, the frequency response of the healthy human ear, peaking at 1000–2000 Hz, and masking level functions. Louder notes tend to obscure weaker notes in both frequency and time, making the ordering of complex tones more complicated.

This work presents the mean frequency, and moments about the mean, of a set of strummed strings as a first approximation. It reduces an entire range of frequencies in a tone or timbre down to a few numbers that a micro-controller might use to order the switching of pickup circuits. Equation 3.4 analyzes the results obtained from strumming six strings of the dual-humbucker electric guitar in the experiment several times. The guitar output signal fed into a desktop computer microphone input and was analyzed by a shareware FFT program.[2] The program produced a set of real number magnitudes (no phase information) in frequency bins, measured in decibels of full scale (dBFS). "Full scale" being the full range of the input of the computer audio board.

$$\mathrm{lin}V_n(f_n) = 10^{\mathrm{dBFS}_n/20}, \quad 1 \le n \le 2048$$

$$P_V(f_n) = \frac{\mathrm{lin}V_n}{\displaystyle\sum_{n=1}^{2048} \mathrm{lin}V_n}$$

$$\mathrm{Mean} \cdot f = \sum_{n=1}^{2048} f_n * P_V(f_n) \tag{3.4}$$

$$2\mathrm{nd} \cdot \mathrm{moment} \cdot f = \sum_{n=1}^{2048} (f_n - \mathrm{mean} \cdot f)^2 * P_V(f_n)$$

$$3\mathrm{rd} \cdot \mathrm{moment} \cdot f = \sum_{n=1}^{2048} (f_n - \mathrm{mean} \cdot f)^3 * P_V(f_n)$$

The program was set to analyze the input at 44,100 samples/s, with 4096-sample windows, shaped by a Hann (raised cosine) filter, averaging all the windows together. This produced 2048 frequency bins from 0 to 22,039 Hz, with a resolution of about 10.77 Hz. Six guitar strings were strummed all at once, for one time at fret, from fret 0 to fret 5.

---

[1]French, R. M. (2009). *Engineering the guitar: Theory and practice.* New York: Springer. Retrieved from https://www.springer.com/us/book/9780387743684.

[2]Steer, W. A. (2001–2016). SpecAn_3v97c.exe, Simple Audio Spectrum Analyzer v3.9, ©W.A. Steer 2001–2016. Retrieved from http://www.techmind.org/audio/specanaly.html

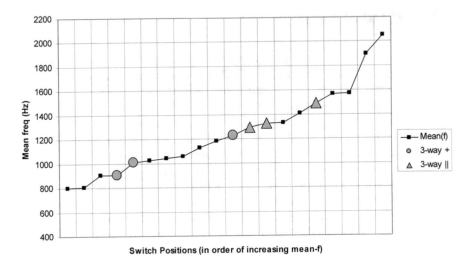

Fig. 3.9 Results of a 24-way switching circuit (with 20 different pickup circuit topologies) for 2 mini-humbuckers on an electric guitar. Mean frequencies of six strings, strummed successively at frets 0–5, sampled at 44.1k-samples/s, with Hann (raised cosine) windows of 4096 samples, converted by FFT to spectral amplitudes in bins 10.77 Hz wide, averaged over all windows taken. Small black squares are mean frequency, the first-order moment of the averages spectral amplitudes. Gray circles are the equivalent circuits for a 3-way switch with internal humbucker coils connected in series. Gray triangles are equivalents of a 3-way switch with parallel-connected coils. From http://TulsaSoundGuitars.com/tutorials/preliminary-results-of-patent-pending-20-way-dual-humbucker-switching/ and Fig. 5, Baker (2017a)

Figure 3.9 shows the results of an experimental test of a prototype guitar with two Hofner-style mini-humbuckers and a 20-way switching network, with no capacitors in Fig. 3.7. The humbuckers were mounted as near as possible to the neck and bridge of a modified electric guitar. At each switch position, all six strings were slowly strummed 6 times, midway between the humbuckers. The first strum was unfretted, fret 0; the second on fret 1, successively up to fret 5. This was done to produce a wider and smoother range of spectral output.

A desktop computer microphone input received the guitar output. FFT software, SpecAn_3v97c.exe, Simple Audio Spectrum Analyzer v3.9 ©W.A. Steer 2001–2016, accumulated the audio data and produced an FFT amplitude spectrum. The software was set to: Amplitude scale = 135 dBFS; zero-weighted; Freq scale = log; Visualize = Spectrograph w/avg; Sample rate = 44.1 kHz; FFT size = 4096; FFT window = Hann cosine. The audio volume pot on the guitar was set to avoid clipping. Each FFT spectral average was exported to a text file with a ∗.csv suffix filename, then imported into an MS Excel spreadsheet.

This produced 2048 frequency "buckets," from 0 to 21,039 Hz, with an average value in dB for each amplitude and a frequency resolution of about 10.7 Hz. This may be why the mean·frequencies shown in Figure 3.9 are so high, far above the string fundamentals. Later experiments used lower sample rates with lower bandwidths and higher resolution in frequency, which did produce lower mean·f results. Sample rates of 16,000/s and 8000/s dropped one mean frequency number from 1170 to 605.6 Hz and 493.3 Hz, respectively.

Strumming over six frets to produce as wide as possible a range of frequencies, with a frequency resolution of 10.7 Hz, also broadens any spectral peaks of fundamentals and harmonics. If the frequency resolution had instead been 1 Hz, and only one string strummed on one fret, the spectral peaks would have been sharper and the mean frequencies much lower, as confirmed by later experiment. So the results can only be taken to demonstrate that the 20-way switching circuit has relatively wider range and finer tonal distinctions than a 3-way switching circuit.

The 2048 buckets and their amplitudes were imported into a spreadsheet. Equation 3.4 shows how the data was processed to obtain the 1st, 2nd, and 3rd frequency distribution moments. The average spectral amplitude was converted from log to linear voltage, linV_n, n going from 1 to 2048. From this a frequency spectral density function, $P_V(f_n)$ was constructed by dividing each value of lin$V$ by the sum of the values. The 1st moment, or mean frequency, mean·f, was then the sum of the product of f_n times $P_V(f_n)$. The second moment is the sum of the product of $(f_n - \text{mean} \cdot f)^2$ times the spectral density. And the 3rd moment is the sum of the cube of $(f_n - \text{mean} \cdot f)$ times the spectral density. Only mean·f is plotted in Fig. 3.9, ordered by increasing mean·f.

The results with $C1 = 3.3$ nF and $C2 = 10$ nF are not plotted, but all the mean frequencies were significantly lower and had a smaller range. With the capacitors in the circuit, the two lower circle and triangle 3-way switch comparisons all got shoved together and below 1000 Hz. It did not seem a useful comparison. More work on the use of resonant peaking capacitors needs to be done.

The differences in frequency between adjacent values of mean frequency run from 0.44 to 326.5 Hz, with an average difference of 65.9 Hz and a standard deviation of 75.8. The smallest differences, less than the resolution of the FFT, occur at 7.5 Hz between points 1 and 2, 0.44 Hz between 3 and 4, 9.0 Hz between 13 and 14, and 0.51 Hz between 17 and 18. The three largest differences are 102.1 Hz between points 4 and 5, 326.5 Hz between 18 and 19, and 154.0 Hz between 19 and 20. Removing these largest and smallest differences changes the mean difference to 54.3 Hz with a standard deviation of 29.5. If one removes the 4 points with the smallest difference to the one above it, one could argue that there are only 16 effectively different tones, out of a 20-way switching system with 24 switch positions. But at least they are all humbucking.

Six data points in the plot correspond to a 3-way switch, 3 designated by circles for 2 humbuckers with their internal coils connected together in series, and 3 by triangles for 2 humbuckers with their internal coils connected together in parallel. In some guitars, the bridge humbucker may be "hotter," with a stronger signal output, than the neck humbucker, to compensate for the smaller relative motion of the strings near the bridge. The humbuckers used here had equal outputs, which may account for the bunching together of two circles and two triangles in the lower range of each, but that is not conclusive. More research is needed.

Note: In the case of humbuckers, since individual coils within each humbucker are matched in turns to each other, the number of turns from humbucker to humbucker do not have to be matched in the KK_{JJT} combinations shown in Table 3.1 for the whole circuit to remain humbucking. The practical limits for how many pickups, single-coil or humbucker, can be placed along the strings is limited for most electric guitars, but not pianos, to the space between the bridge and neck. Besides which, the closer individual coils come to each other in space, the more their fields interact, and transfer vibrational energy between them, causing tonal and phase effects which cannot be addressed here.

As well as the zero-weighting used in the experiment, the SpecAn program also offered A-weighting, which this author did not understand at the time, so did not use. As French (2009) noted, A-weighting corresponds to the frequency response of the healthy human ear, peaking above 1000 Hz, and falling off for lower and higher frequencies. Zero-weighting represents how a microphone with a uniform frequency response would present the signal. If the spectra had been A-weighted, the mean frequencies could have come out considerably different. According to Wikipedia,[3] the official equation for A-weighting[4] (IEC, 2013) is shown below in Eq. (3.5), which shows a peak at about 2508 Hz.

$$R_A(f) = \frac{12,194^2 \cdot f^4}{(f^2 + 20.6^2)\sqrt{(f^2 + 107.7^2)(f^2 + 737.9^2)}(f^2 + 12,194^2)} \quad (3.5)$$

$$A(f) = 20\log_{10}(R_A(f)) + 2.00$$

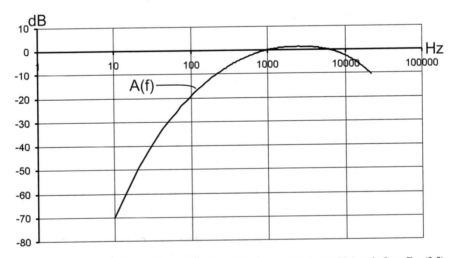

Fig. 3.10 The EC 61672 A-weighting function, $A(f)$, in log (dB)-linear (Hz) scale from Eq. (3.5) for the range of spectral frequencies from the experiment in Fig. 3.9

---

[3]https://en.wikipedia.org/wiki/A-weighting
[4]IEC. (2013). EC 61672-1:2013 Electroacoustics—Sound level meters—Part 1: Specifications. Retrieved from https://webstore.iec.ch/publication/5708.

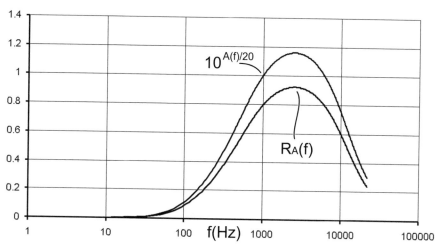

Fig. 3.11 $R_A(f)$ from Eq. (3.5), and linearized $A(f)$, for the range of frequencies in spectra from the experiment

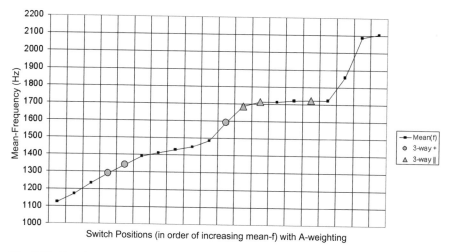

Fig. 3.12 Experiment data from Fig. 3.9, modified by multiplying the linearized A-weighting function times $\lin V_n(f_n)$ before calculating the mean frequency

Figures 3.10 and 3.11 show the functions from Eq. (3.5), with $A(f)$ linearized in Fig. 3.11. $R_A(f)$ has a peak of 0.9195 at 2508 Hz, and the linearized $A(f)$ has a peak of 1.1576. Figure 3.12 shows the experimental data, modified by multiplying the linearized $A(f)$ curve in Fig. 3.11 times the $\lin V_n(f_n)$ in Eq. (3.5) before calculating the mean frequency and spectral moments. Note how the range of mean frequency has moved from between 808 to 2052 Hz in Fig. 3.9 to 1125 to 2104 Hz in Fig. 3.12, and how the shape and distribution of the points have changed. Figure 3.9 is what an amplifier "hears," with a mean frequency range of about 2.54 to 1. Figure 3.12 is approximately what a normal human ear perceives, with a mean frequency range of about 1.87 to 1. Nor are the switching positions (not shown) in the same order.

In Fig. 3.12, the frequencies on the shoulder at just above 1700 Hz are close together, with $(1713.1 - 1711.7) = 1.4$ Hz, $(1720.9 - 1713.1) = 7.8$ Hz, $(1722.4 - 1720.0) = 1.4$ Hz, and $(1723.1 - 1722.4) = 0.8$ Hz. Only 1723.1 Hz is enough different from the 1711.7 and 1856.8 Hz to be counted as more different than the 10.77 Hz spectral bin size. So 1713.1, 1720.9, and 1722.4 Hz are too close to other mean frequencies to count, leaving 17 presumably distinct tones.

In later experiments (intended for later chapters) with the same two mini-humbuckers, 8000/s sample rates were used, with a high end frequency of 3998 Hz. The spectra show a distinct drop-off above 1000 Hz, and the mean frequencies with zero-weighting were much lower. So the plots presented here in Figs. 3.9 and 3.12 should be taken only as relative differences between 3-way switches and 20-way switches, and between zero- and A-weighting. This particular switching system, using two 4P2T toggle switches, and a 4P6T rotary switch, cannot be ordered for monotonic changes in the warm or brightness of tone. Switching systems yet to be presented can.

References

Anderson, T. S. (1992). Electromagnetic pickup with flexible magnetic carrier, US Patent 5,168,117, 1 Dec 1992. Retrieved from https://patents.google.com/patent/US5168117A/

Baker, D. L. (2017a). Humbucking switching arrangements and methods for stringed instrument pickups, US Patent Application 15/616,396, filed 7 June 2017, published as US-2018-0357993-A1, 13 Dec 2018, granted as Patent US10,217,450, 26 Feb 2019. Retrieved from https://www.researchgate.net/publication/335727402_Humbucking_switching_arrangements_and_methods_for_stringed_instrument_pickups_-_NPPA_15616396

Blucher, S. L. (1995). Transducer for stringer musical instrument, US Patent 4,501,185, 26 Feb 1985. Retrieved from https://patents.google.com/patent/US4501185A/

Fender, C. L. (1961). Electromagnetic pickup for lute-type musical instrument, US Patent 2,976,755, 28 Mar 1961. Retrieved from https://patents.google.com/patent/US2976755A/

French, R. M. (2009). Engineering the guitar: Theory and practice. New York: Springer. Retrieved from https://www.springer.com/us/book/9780387743684

IEC. (2013). EC 61672-1:2013 Electroacoustics—Sound level meters—Part 1: Specifications. Retrieved from https://webstore.iec.ch/publication/5708

Lesti, A. (1936). Electric translating device for musical instruments, US Patent 2,026,841, 7 Jan 1936. Retrieved from https://patents.google.com/patent/US2026841A/

Lover, S. E. (1959). Magnetic pickup for stringed musical instrument, US Patent 2,896,491, 28 July 1959. Retrieved from https://patents.google.com/patent/US2896491A/

Steer, W. A. (2001–2016). SpecAn_3v97c.exe, Simple Audio Spectrum Analyzer v3.9, ©W.A. Steer 2001–2016. Retrieved from http://www.techmind.org/audio/specanaly.html

Chapter 4
Series-Parallel Circuit Topologies of Matched Single-Coil Pickups

4.1 Some History

In searching patents, this author never found evidence of any other inventor developing matched single-coil pickups in humbucking combinations. With four identical single-coil pickups under a pickguard, two N-up and two S-up, Fender (US3,290,424, 1966) came as close as any. Numbers 27 and 29 were S-up, and numbers 28 and 30 N-up, with 27 at the bridge and 30 at the neck. But he apparently never made the connection to humbucking circuits. Instead, he connected each coil to an all-parallel circuit with a 2P3T switch, which connected the pickup in normal, reversed, and out of the circuit. This presented $3^4 = 81$ different switch configurations, of which one had no output. Call the pickups from the neck to the bridge N1, S1, N2, and S2. The 9 possibly unique humbucking combinations are N1∥S1, N1∥(−N2), N1∥S2, S1∥N2, S1∥(−S2), N2∥S2, N1∥S1∥N2∥S2, N1∥S1∥(−N2)∥(−S2), and N1∥(−S1)∥(−N2)∥(S2). All the rest of the 80 switch combinations with outputs are either inverted duplicates or non-humbucking.

If the middle pickup of a 3-coil guitar has the opposite pole up from the neck and bridge coils, and all three coils have the same number of turns and impedances, then positions 2 and 4 of a 5-way switch have humbucking outputs. But no patent that this author could find has made such a claim. Furthermore, the bridge coil of a 3-coil guitar tends to be made hotter to pick up the smaller vibrations of the strings, thus reducing the possible humbucking properties of connecting it in parallel with the middle coil.

Somewhere during or before April 11, 2011, this author was working on "Switching circuit for 4 single coil pickups acting as a humbucker" (from notes), trying to find the most efficient circuit. Then and again in October 2011, various circuits of 8P5T, 5P5T, 4P5T, 8P4T, 8P6T, 4P7T, 4P4T circuits were drawn and ultimately rejected. Finally on June 16, 2012, a circuit using a 4P10T switch allowed the switching of all humbucking series and parallel outputs merely by connecting the low terminals of the S-up pickups to

© Springer Nature Switzerland AG 2020
D. L. Baker, *Sensor Circuits and Switching for Stringed Instruments*,
https://doi.org/10.1007/978-3-030-23124-8_4

the low output terminals and the high terminals of the N-up pickups to the high output terminal. This later became Figs. 27–29 in US 9,401,134 B2,[1] as shown below in Figs. 4.1, 4.2, and 4.3.

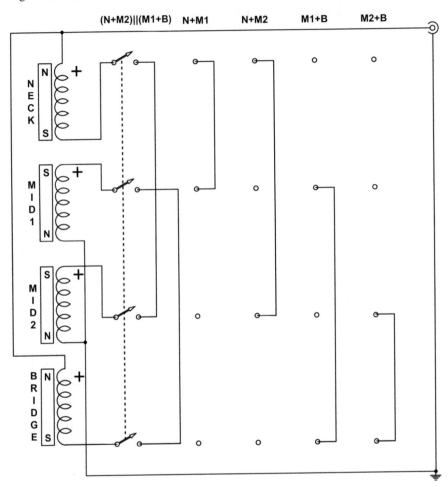

Fig. 4.1 All-humbucking 4-pole 5-throw series-connected switch combinations for four matched single-coil pickups with two N-up and two S-up poles. The "+" signs on the coils indicate the positive vibration signal phases. From Fig. 27, Baker (2016b)

Figure 4.1 shows a 4P5T super-switch with four in-phase series connections and one series-parallel connection for four matched single-coil pickups. A super-switch has two switching wafers with two poles per wafer. It is at least twice as wide and a bit deeper than a standard 5-way guitar switch. Notice that the positive phase

---

[1]Baker, D. L. (2016b). Acoustic-electric stringed instrument with improved body, electric pickup placement, pickup switching and electronic circuit, US Patent 9,401,134, 26 July 2016. Retrieved from https://patents.google.com/patent/US9401134B2/.

terminals of the N-up pickups are connected to the high terminal of the output, and the negative phase terminals of the S-up pickups are connected to the low terminal, or ground. With this setup and a limited number of poles, parallel connections can be made, as in Fig. 4.2, but no out-of-phase circuits can be constructed.

However, call the neck, middle1, middle2, and bridge pickups N1, S1, S2, and N2, respectively. And let "+" and "−" denote series connections, and "‖" denote a parallel connection. If the terminals of N1 and S1 are reversed to (−N1) and (−S1), then these series humbucking connections can be made: −(N1 + S1), (N2 − N1)-out-of-phase, (S2 − S1)-out-of-phase, (N2 + S2), and either (N2 − N1)‖(S2 − S1) or −(N1 + S1)‖(N2 + S2). If the terminals of S1 and N2 are reverse, then the possible series humbucking connections include: (N1 + S2), (N1 − N2), (S2 − S1), −(N2 + S1), (N1 − N2)‖(S2 − S1), and −(S1 + N2)‖(N1 + S2). There is no way in this switching setup to connect all four in series. This illustrates the limits of electromechanical switches, where connections are hard-soldered.

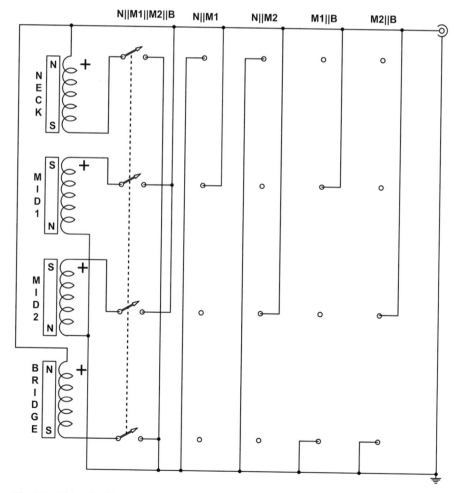

Fig. 4.2 All-humbucking 4-pole 5-throw parallel-connected switch combinations for four matched single-coil pickups with two N-up and two S-up poles. The "+" signs on the coils indicate the positive vibration signal phases. From Fig. 28, Baker (2016b)

Figure 4.2 follows Fig. 4.1, except that the connections are parallel. Again, if the terminals of one N-up and one S-up pickup are reversed, similar out-of-phase signals can be constructed, but in parallel. There is, however, nothing that prevents the position order of the pickups from the neck to the bridge from being changed. It could just as well be N1, N2, S1, S2. And in US 9,401,134, the positional placement of the pickups can be changed in three dimensions with 5 degrees of freedom, up and down under the strings at either end of the pickup, toward or away from the neck at either end, and movement across the strings. In other words, position and cant between the neck and bridge, tilt and position under the strings, and position across the strings.

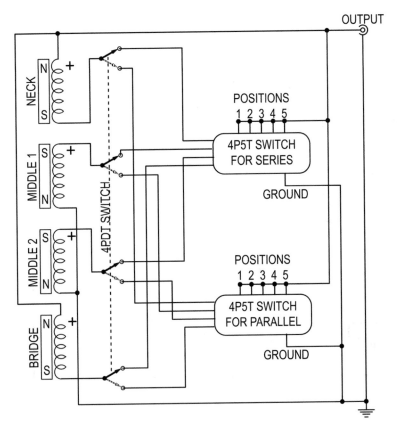

Fig. 4.3 A 10-way switching system for four matched single-coil pickups using a 4P2T switch and two 4P5T "super-switches" from Figs. 4.1 and 4.2. From Fig. 29, Baker (2016b)

Figure 4.3 shows how the 4P5T switches in Figs. 4.1 and 4.2 are combined into a 10-way switching system. The tone and volume circuits are not shown.

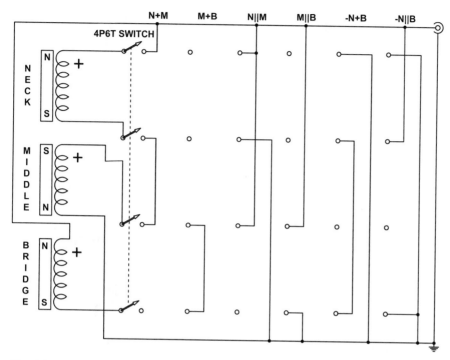

Fig. 4.4 A 6-way switching system using a 4P6T rotary switch, for three matched single-coil pickups, one S-up (middle) and two N-up, with the "+" signs noting the positive phase of the string vibration signals. "+" and "−" denote series connections and "‖" denotes parallel. From Fig. 24, Baker (2016b)

Figure 4.4 shows a 6-way switching system for three single-coil pickups, which first appeared in the inventor's notes on or before January 12, 2014. Notice that the neck pickup uses 2 switch poles, so that its terminals could be reversed. This circuit, leaving out the –N‖B connections, was wired on a 5-way switch[2] (with options for other connections[3]) and was wired on a standard 3-coil guitar,[4] shown in Fig. 4.5. One professional guitarist allowed that it sounded pretty good, and seemed to like the contra-phase –N + B position.

---

[2]Baker, D. L. (2016). An all-humbucking 5-way superswitch circuit. http://android-originals.com/tulsasound/tutorials/an-all-humbucking-5-way-superswitch-circuit/

[3]Baker, D. L. (2016). HB supersw circuit for 3 p/u with odd pole at neck. http://android-originals.com/tulsasound/tutorials/hb-supersw-circuit-for-3-pu-w-odd-pole-at-neck/

[4]Baker, D. L. (2006, 2007). First guitar with art decal. http://theandroidsaxe.com/Decals/index.html

Fig. 4.5 A 3-coil guitar decorated with vinyl van wrap material. Eventually used to test a 5-way all-humbucking superswitch circuit. http://theandroidsaxe.com/Decals/index.html

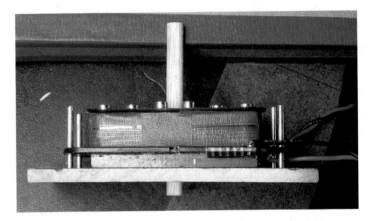

Fig. 4.6 One of 2 generic single-coil pickups to which turns have been added, over masking tape covering original coil, mounted to Masonite rectangle on brass pins. Another piece of blue masking tape secures the extra turns. A central wood dowel post sits in a hole, centered in the slanted piece of brown Masonite seen beneath it

Fig. 4.7 Test setup for two generic single-coil pickups, made of Masonite (bottom strip for centering pickups), cardboard, wood dowel, brass pins, popsicle sticks, glue, masking tape, and several hundred turns of 30 gage wire-wrap wire. The blue-painted pickup is N-up, and the red is S-up. The S-up poles had to be reversed from N-up by wiping with a much stronger rare-earth magnet

Figures 4.6 and 4.7 show a test setup for checking if enough turns had been added to match two generic single-coil pickups. The coil measures about 180×254 mm (7×10 in.). The coil was made of #30 wire-wrap wire, and the number of turns has been forgotten. It was probably between 100 and 200. A Simpson Model 420 function generator drove it with a 60–100 Hz sine wave, to a level which produced about 950 mVpp on the output of each pickup.

The pickups were tested together and individually to determine the order of sensitivity to the hum coming from the coil. Then the rest of the coils were matched as closely as possible to the one with the strongest hum signal by adding turns of #37 magnet wire, using blue masking tape to insulate solder connections and new turns. For the matching comparison, two coils were connected in series, so that the outputs opposed each other. Then turns were added manually to reduce the contra-phase hum voltage to a few millivolts. This made the turns a bit loose, which may have contributed to a "reverb" quality of their output when placed in a prototype guitar. Either that or the semi-cantilevered sound board.

Three of the generic single-coil pickups used had a coil resistance of about 7.75 kΩ, and one had a coil resistance of 5.2 kΩ, with a different size of original wire from the others. On an oscilloscope, the contra-phase hum signal showed a phase difference, indicating an different coil impedance, when this coil was connected in series with the others. This emphasizes that matched pickups must be made exactly the same way, with the same wire and number of turns.

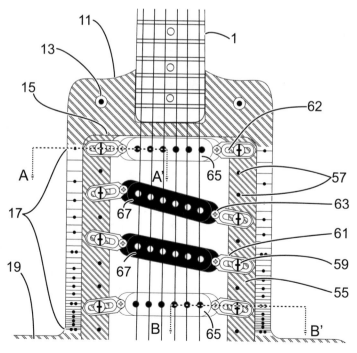

Fig. 4.8 A portion of Fig. 1 from US 9,401,134 B2 (Baker, 2016b), showing single-coil matched pickup mounting that allows movement in 5 degrees of freedom

Figure 4.8 shows a portion of Fig. 1 from US 9,401,134 B2 (Baker, 2016b), with a neck (11), skeletonized body (11), mounting posts for the sound board (13), an extended virtual fret scale (17), a mounting rail for pickups (55) with threaded holes (57) for mounting screws (59), which fix pickup mounting plates (61) to the body. Slots (62) in the plates allow them to slide and rotate, placing the matched single-coil pickups (65, N-up and 67, S-up) at any angle and position under the strings, at any position between the neck and bridge, using the mounting holes and screws. Pickup mounting screws (63, with springs not shown) set the high of the ends of the pickups under the strings.

For adjusting this top-mounting configuration, a tunematic style bridge was mounted to an aluminum flat bar, in turn mounted to the body in the bridge position, in place of the sound board. This left the most of the top open. The regular sound board, a flat plate with 5-point mounting and free-floating edges, covered the pickups entirely. Of course, once the pickups have been placed to produce a desire tonality for the guitar, they can be mounted directly in the sound board in the same positions. A better variation puts the mounting plates on the underside of the body, so the pickups can be adjusted by removing the back plate of the body, without disturbing the bridge or sound board. For more details, see the patent.

4.2   Humbucking Pairs, Quads, Hextets, and Octets

What makes a pair of matched pickups humbucking? By definition, they have the same coil impedance and the same response to external hum. If they are connected, either in series or parallel so that the hum voltages oppose in contra-phase, the hum voltages effectively cancel each other out. As such, it does not matter if both have the same pole up or opposite poles up toward the strings. The poles only determine if the string vibration signals will be in-phase or contra-phase. If the poles are aligned in the same direction, the string signal will be contra-phase. If they are aligned in opposite directions, the string signals will be in-phase. For simple series-parallel circuits, those are the only two valid humbucking options.

When first considered by this author, Fig. 3.1 also began the explanation of the conversion of single-coil circuit topologies in Figs. 2.4–2.9 into humbucking pair topologies, and up, of which dual-coil humbuckers are a sub-set. But with matched single-coil pickups, it is easier than with dual-coil humbuckers to pass a single matched coil around to different positions in a given circuit topology. This can be done with dual-coil humbuckers, but only if the coils have all four coil terminals coming out as signal wires. Many if not most humbuckers are sold as 2-wire pickups, with the coils connected in series internally, possibly so that no mistakes are made by users in connecting them together as humbucking. The mini-humbuckers in Figs. 3.2 and 3.3 had to be opened up and rewired with 4-wire shielded cable, which turned out to be a bit stiff.

4.2.1 Humbucking Pairs

It's useful for later to start out with some circuits and circuit equations. Figure 4.9 below shows a circuit representation of two matched single-coil pickups, one N-up on the left with string signal voltage V_N, and one S-up on the right with string signal voltage V_S. The impedances, Z, and hum voltages, V_H, are matched and equal. The sign conventions are just that, conventions set up for this discussion. It would work equally well if the signs of the signal voltages or the signs of the hum voltages were reversed.

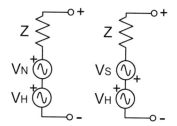

Fig. 4.9 Two matched single-coil pickups, represented as voltages and equal impedances Z, with North-up signal voltage V_N on the left, South-up signal voltage V_S on the right, and equivalent hum voltages V_H in both. From Fig. 1, Baker (2018b)

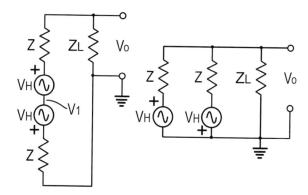

Fig. 4.10 Two matched pickups in series (left) and parallel (right) connections, with no string signals. The triangle with stripes indicates the common ground connection, to which circuit voltages are referenced. From Fig. 4a, b, Baker (2018c)

Figure 4.10 shows two matched pickups with only the hum signals showing. We can say that the strings are quiet, but in linear circuit analysis, voltages sources can be considered one at a time and added at the output. The left circuit is a humbucking series connection, with voltage V_1 at the connection, and the right is a humbucking parallel connection. Both circuits have a load impedance, Z_L. Almost any symbolic math software package can be used to solve the circuit equations, but this author happens to use an old student copy of Maple V.

This book generally assumes that the reader has some training and understanding in electronic circuits and circuit theory. But for those who may not, here is a little tutorial. This kind of linear circuit equation, in the listings above, is based upon the principle that all the current flowing into a voltage node must flow out of it. Current flowing into reactive components is another matter, which is mostly outside the scope of this book. Current is equal to voltage over resistance, or impedance. Also all the voltage sources and impedances in series between two nodes can simply be summed together. Therefore, V_1 in Fig. 4.10 and Listing 4.1 was not strictly necessary. Listing 4.3 shows the simpler version of Listing 4.1.

Listing 4.1 shows the Maple solution for the left circuit, and Listing 4.2 shows the solution for the right circuit. In both solutions, Vo = 0. By inspection, the impedance of the series connection is 2Z, and of the parallel connection, Z/2.

Listing 4.1 Maple V Solution of Circuit Equations for Series Connection in Fig. 4.10

```
STUDENT > restart; # series circuit
STUDENT > e1 := (v1+vh)/z + (v1+vh-vo)/z = 0;
```

$$e1 := \frac{v1 + vh}{z} + \frac{v1 + vh - vo}{z} = 0$$

```
STUDENT > e2 := (vo-v1-vh)/z + vo/zl = 0;
```

$$e2 := \frac{vo - v1 - vh}{z} + \frac{vo}{zl} = 0$$

```
STUDENT > solve({e1,e2},{v1,vo});
```

$$\{ v1 = -vh,\ vo = 0 \}$$

> restart; # series circuit
> e1 := (v1+vh)/z + (v1+vh-vo)/z = 0;

$$e1 := \frac{v1 + vh}{z} + \frac{v1 + vh - vo}{z} = 0;$$

> e2 := (vo-v1-vh)/z + vo/zl = 0;

$$e2 := \frac{vo - v1 - vh}{z} + \frac{vo}{zl} = 0;$$

> solve({e1,e2},{v1,vo});

$$\{ v1 = -vh,\ vo = 0 \}$$

Listing 4.2 Maple V Solution of Circuit Equation for Parallel Connection in Fig. 4.10

```
STUDENT > restart; # parallel connection
STUDENT > e1 := (vo-vh)/z + (vo+vh)/z + vo/zl = 0;
```

$$e1 := \frac{vo - vh}{z} + \frac{vo + vh}{z} + \frac{vo}{zl} = 0$$

```
STUDENT > solve(e1,vo);
```

$$0$$

> restart; # parallel connection
> e1 := (vo-vh)/z + (vo+vh)/z + vo/zl = 0;

$$e1 := \frac{vo - vh}{z} + \frac{vo + vh}{z} + \frac{vo}{zl} = 0;$$

> solve(e1,vo);

$$0$$

Listing 4.3 Alternative Maple V Solution of Circuit Equation for Series Connection in Fig. 4.10

>restart; #series HB pair
>e1 := (vo-vh+vh)/2/z + vo/zl = 0;

$$e1 := \frac{1}{2}\frac{vo}{z} + \frac{vo}{zl} = 0;$$

>solve(e1,vo);

$$0$$

In each term of the equation, the difference between the voltage at the target node and the voltage at the secondary node is divided by the resistance or impedance between the nodes. The target node is considered "high" and the secondary node is considered "low." There must be an equation for every node, except ground, indicated by the striped triangles in Fig. 4.10, which is always considered at zero voltage. Source voltages between the nodes, such as the V_H hum voltages, indicated by the symbol, vh, in the listings, are added or subtracted as they aid to or oppose the voltage of the primary node.

The load impedance, Z_L, or the alternative load resistance, R_L, are not always strictly necessary. They may fall out entirely in the solution. But they often insure that no coefficients are dropped by mistake. In this case, if the load resistance or impedance shows in the solution, the solution is taken to the limit of Z_L or R_L equal to infinity, to obtain the no-load output of the pickup circuit. This is the voltage that would appear at the output, Vo, from hum or string vibration signals.

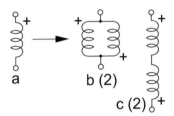

Fig. 4.11 Replacing a single-coil pickup (**a**) with a humbucking pair of matched single-coil pickups (**b**, **c**) with basic topologies of type (2). The "+" signs indicate the polarities of the hum voltages. From Fig. 4, Baker (2017a)

Figure 4.11 shows the basic manner for converting the topologies of unmatched single-coil pickups in Figs. 2.4–2.9 to humbucking pairs, quads, and up. In this way a single coil becomes two possible humbucking pairs. Circuits of 2 coils become humbucking quads, circuits of 3 coils become humbucking hextets, and so on. Note in Fig. 4.11 with just two coils, the choices are pretty limited. Exchanging the two coils with each other can't even change the signal phase, and a single coil cannot be reversed without destroying humbucking.

Until someone can prove differently, this author contends that merely reversing the output terminals of either pair b or c cannot produce a new tone. It can only invert the signal, which phase this author contends that the human ear cannot detect without another phase reference. This argument rests upon the fact that the only difference between $\sin(\omega t)$ and $-\sin(\omega t)$ is the time period of a half cycle, which gets increasingly smaller with increasing frequency, $\omega = 2\pi f$, where f is frequency in Hz. Call it the *Principle of Inverted Duplicates*. We will see it again.

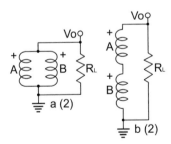

Fig. 4.12 Parallel, a(2), and series, b(2), pair of matched single-coil pickups, A and B, of basic topology type (2), with signal polarities shown by "+," equal impedance, Z, output Vo and load R_L

Now consider Fig. 4.12, where A and B can stand in for two matched pickups, their string signal voltages, or their hum voltages, as needed. Both have impedances, Z. Listing 4.4 shows the Maple V solution to the output signal, Vo, and the circuit lumped impedance, Zc. The first equation, e1, is the circuit equation for each pair, just like the Listings above. The second equation, e2, depends upon the fact that the load is equal to the lumped circuit impedance, the output is ½ of the output with no load, as shown in Eq. (4.1). So, solve the equation e2 for R_L, and get the value of Zc in terms of Z, Z/2 for the parallel circuit, and 2Z for the series circuit. This will be useful when the circuits become more complex.

$$\frac{1}{2} Vo|_{R_L=\infty} = Vo|_{R_L=Z_C} \tag{4.1}$$

Listing 4.4

restart; #parallel pair
>e1 := (vo-a)/z + (vo-b)/z + vo/rl = 0;

$$e1 := \frac{vo - a}{z} + \frac{vo - b}{z} + \frac{vo}{rl} = 0$$

> solve(e1,vo);

$$\frac{(a + b) rl}{2 rl + z}$$

> limit((a+b)*rl/(2*rl+z), rl=infinity);

$$\frac{a + b}{2}$$

> e2 := (a+b)/4 = (a+b)*rl/(2*rl+z);

$$e2 := \frac{a + b}{4} = \frac{(a + b) rl}{2 rl + z}$$

> solve(e2,rl);

$$\frac{1}{2} z$$

> restart; #series pair
> e1 := (vo-a-b)/2/z + vo/rl = 0;

$$e1 := \frac{vo - a - b}{2 z} + \frac{vo}{rl} = 0$$

> solve(e1,vo);

$$\frac{(a+b)\,rl}{rl+2\,z}$$

> limit((a+b)*rl/(rl+2*z), rl=infinity);

$$a+b$$

> e2 := (a+b)/2 = (a+b)*rl/(rl+2*z);

$$e2 := \frac{a+b}{2} = \frac{(a+b)\,rl}{rl+2\,z}$$

> solve(e2,rl);

$$2\,z$$

Now consider that both solutions for Vo are factors of the combined signal voltage (A + B). For a no-load situation, that is the same tone, just at different amplitudes. The only difference in tone between a series and parallel humbucking pair occurs when they are connected to the same value of load, and that load is close enough to the value of Z to form a low-pass filter, which cuts off higher frequencies above a certain level. Say the load is a 250k–500k volume pot. That load looks about 4 times smaller to a circuit with an impedance of 2Z than it does to a circuit with an impedance of Z/2. Since the lumped impedance of a pickup circuit goes up with frequency, the load looks even smaller as the frequency of the signal increases.

So the series circuit with a lumped impedance of 2Z loses more of the higher frequency harmonics to the voltage divider effect of the load than a parallel circuit with a lumped impedance of Z/2. That is why in a guitar circuit with passive components, series-connected pickups sound warmer than when the same pickups are connected in parallel. When the pickup circuit is connected to a high-impedance input of an active amplifier, that difference occurs only at frequencies tending to be higher than can be heard. That tonal difference effect goes away. So the series- and parallel-connected humbucking pairs tend to sound different only in amplitude, not tone. If the pickup circuits are amplified, the differences in tone will come from differences in the proportions of pickup signals that appear at the output—as we will see later. The variety of tones will appear to decrease. We can call these "no-load tones," as opposed to "loaded tones," which can be affected directly by the passive attachment of tone and volume controls directly to the pickup circuit output.

Table 4.1 shows the solutions for Vo in Fig. 4.12, the lumped circuit impedance with no load, and the relative signal intensity for Vo, if A and B are essentially the same signals. In this case, for convenience, we are assuming that A and B are in-phase, because one pickup is N-up and the other is S-up, putting the hum signals out-of-phase. As in Fig. 4.10, humbucking depends not upon the relative direction of the magnetic fields, but upon the relative hum phase of the coils. So long as a circuit is humbucking by that standard, it does not matter which pickup has what magnetic pole up. Two of the same poles up will automatically be contra-phase, and two pickups of different poles will automatically be in-phase.

Table 4.1   Solutions for Vo for circuit topologies of Fig. 4.12, with relative intensities taking A and B as equal signals

Figure 4.12 circuit	Solution for Vo	Lumped impedance	Relative intensity
a(2)	(A + B)/2	Z/2	1
b(2)	A + B	2Z	2

A later chapter will present matched single-coil pickups with magnets that can be manually reversed, allowing a guitar with K number of pickups to have 2^{K-1} number of tonal characters, each having $K * (K - 1)$ humbucking pair tones from a total of $2 * K * (K - 1)$. But this is based upon series-parallel pair connections. Table 4.1 shows that this number can be cut down by at least half, if only no-load tones are available.

How many loaded tones can you get from a humbucking pair? Let the number of humbucking pairs be JP $= 1$. Not by coincidence, the number of terminal reversals of pairs that will produce new phase combinations at the ultimate output is $N_{SGN-HP} = 2^{JP-1} = 2^0 = 1$. The total number of loaded-tone topologies for JP $= 1$ is $N_{LT-HP} = 2$. Suppose that we have K number of matched single-coil pickups. How many humbucking pairs, K_{CP}, of the circuits b and c in Fig. 4.11 can they produce? Equation (4.2) shows that it is the number of combinations of K things taken 2 at a time, times the number of unique topologies, 2, times the number of possible terminal reversals of the pair, 1.

$$K_{CP} = \binom{K}{2} * N_{LT-HP} * N_{SGN-HP} = \frac{K * (K - 1)}{2 * 1} * 2 * 2^{JP-1}$$

$$= \frac{K * (K - 1)}{2 * 1} * 2 * 1 = K * (K - 1), \quad K \geq 2 \tag{4.2}$$

i.e., for $K = 2, 3, 4, 5, 6, 7, 8$; $K_{CP} = 2, 6, 12, 20, 30, 42, 56$

How many no-load tones can you get from a humbucking pair? Instead of $N_{LT-HP} = 2$, we use $N_{NL-HP} = 1$, and get Eq. (4.3).

$$K_{\text{NL-CP}} = \binom{K}{2} * N_{\text{NL-HP}} * N_{\text{SGN-HP}} = \frac{K * (K - 1)}{2 * 1} * 1 * 2^{\text{JP}-1}$$

$$= \frac{K * (K - 1)}{2 * 1} * 1 * 1 = K * (K - 1)/2, \quad K \geq 2 \tag{4.3}$$

i.e., for $K = 2, 3, 4, 5, 6, 7, 8;$ $K_{\text{NL-CP}} = 2, 3, 6, 10, 15, 21, 28$

4.2.2   Humbucking Quads

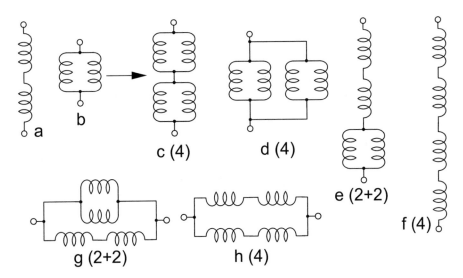

Fig. 4.13 Conversion from $J = 2$ unmatched single-coil topologies in Fig. 2.7b to humbucking pair topologies, by replacing single coils with series and parallel pairs. Numbers in parentheses indicate the basic topologies present in the circuit. The c(4) and h(4) topologies, which look like (2 + 2) topologies, are explained below. From Fig. 6, Baker (2017a)

Figure 4.13 shows the conversion of unmatched single-coil topologies for $J = 2$ to humbucking (HB) pair topologies for JP = 2. For example, in 4.13a, two parallel-connected HB pairs replace the single coils to form topology c(4), which will be shown to have a basic topology of size 4. The same goes for converting 4.13b to d(4). In 4.13e(2 + 2), a series-connected HB pair and a parallel-connected HB pair each replace a single coil in 4.13a. Note that there are no duplicates with obvious horizontal or vertical exchanges, for example, anything corresponding to 4.13e with a parallel-connected pair above a series-connected pair. With the same pickups in the same pairs, the signal across the topology would not change merely by reversing the order of the pairs. So 4.13a converts to c, e and f, and 4.13b converts to d, g and h.

4.2.2.1 HB Quad Circuit Equations

Now, what can circuit equations tell us? Suppose that the pickups in a quad topology are labeled A, B, C, and D, with the hum or signal voltages wired positive toward the output terminal, Vo, and a load resistor, R_L, is connected between Vo and ground. The labels A–D can also be used for hum or string signal voltages, as needed, which will be explained. From this we get Fig. 4.14, where the "+" signs indicate the polarity of the signal voltages. This will also explain why a topology, like Fig. 4.13c (4), which looks like a (2 + 2) topology, is really a (4).

Figure 4.14 shows the same topologies with the coils identified as A, B, C, and D, with "+" signs to indicate initial signal phase. One end of the topology is grounded and the other labeled as the output voltage, Vo, between which a load resistance, R_L, is added. Note that 4.13c, e require an additional node voltage, V1, for use in circuit equations. Listing 4.5 shows the Maple V solution of the circuit equations for 4.13c (4), excluding the solution for V1.

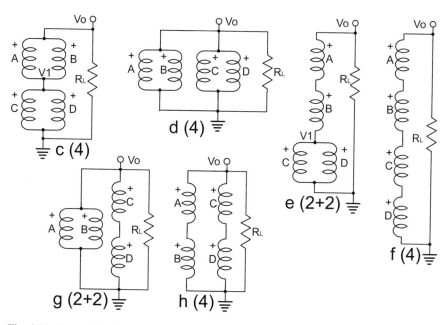

Fig. 4.14 Figure 4.13c–h with matched pickup coils identified and labeled with phase, and with load resistance, ground and output, Vo, added for circuit equations

Listing 4.5 Maple V Circuit Equations for Fig. 4.14c(4), Excluding the Solution for V1

>restart; #c(4)
>e1 := (vo-v1-a)/z + (vo-v1-b)/z + vo/rl= 0;

$$e1 := \frac{vo - v1 - a}{z} + \frac{vo - v1 - b}{z} + \frac{vo}{rl} = 0;$$

> e2 := (v1+a-vo)/z + (v1+b-vo)/z + (v1-c)/z + (v1-d)/z = 0;

$$e1 := \frac{v1 + a - vo}{z} + \frac{v1 + b - vo}{z} + \frac{v1 - c}{z} + \frac{v1 - d}{z} = 0;$$

> solve({e1,e2},{vo,v1});

$$vo = \frac{1}{2} \frac{rl\,(a + b + c + d)}{rl + z}$$

Note that as R_L (shown as rl in Listing 4.4) goes to infinity, Z becomes insignificant, and the solution for the open-circuit voltage becomes Voc = (A + B + C + D)/2. The circuit equation solutions for the rest of the circuit topologies in Fig. 4.14 are left as an exercise for the reader. Table 4.2 shows the solutions for Vo in each case.

Table 4.2 Solutions for Voc (open-circuit voltage) for circuit topologies of Fig. 4.14, with relative intensities taking A, B, C, and D as signals of the same strength and phase

Figure 4.14 circuit	Solution for Voc	Lumped impedance	Relative intensity
c(4)	(A + B + C + D)/2	Z	2
d(4)	(A + B + C + D)/4	Z/4	1
f(4)	A + B + C + D	4Z	4
h(4)	(A + B + C + D)/2	Z	2
e(2 + 2)	[2(A + B) + (C + D)]/2	3Z/2	3
g(2 + 2)	[(A + B) + 2(C + D)]/5	2Z/5	6/5

First we can see that there can't be quite as many different tones as we may have thought. If we take A, B, C, and D to be string vibration signals, then circuits c and h have the same lumped impedance and output, and d and f have the same tone as c and h, but at different amplitudes and impedances. Note that in circuits c, d, f, and h, the order of A, B, C, and D in the expression does not matter. Recall that a basic topology is one between two circuit nodes where changing the physical order of the pickups does not change the tone. These two facts are equivalent. That makes all of these a type (4) topology, with the same no-load tone.

Also note that the difference between the c(4) and h(4) circuits is a single connection between the series circuits on both sides of h(4). This demonstrates a symmetry property in some circuits with even numbers of pickups. When the circuit diagram is flipped either vertically or horizontally, the topology still looks the same. Also note that if the connection were another matched pickup in a set of matched pickups, it would contribute equally to both output terminals. A solution of Voc for a 5-pickup circuit like that will show that it matches c(4) and h(4). Table 4.2 confirms the suggestion on these bases that c(4) and h(4) are equivalent circuits, and that the number of unique tone circuits is less than one might have thought.

4.2.2.2 Allowable HB Quad Pickup Terminal Reversals in Type (4) Topologies

For c, d, f, and h, consider A, B, C, and D to be equal hum voltages. Then two of the hum signs in $(A + B + C + D)$ must be opposite. We can see this best by replacing the signs of the signals with a binary "0" for "+" and a binary "1" for "−." Circuits c, d, f, and h started with a binary number of 0000 representing all "+" signs. If we increment this by one bit each time, we get: 0000, 0001, 0010, **0011**, 0100, **0101**, **0110**, 0111, and 1000.

Note that 0111 and 1000 produce signals for Vo and −Vo, which the human ear cannot (until proven otherwise) tell the difference. So we don't consider results above 0111. The bold, underlined numbers, **0011**, **0101**, and **0110** correspond to signals based on $(A + B − C − D)$, $(A − B + C − D)$, and $(A − B − C + D)$. In other words, when A, B, C, and D are considered hum signals, $Vo = A + B − C − D = 0$, and so on. So there are three possible ways to set up humbucking circuits in a type (4) basic topology. This has previously been called "signs & pairs" in one of this author's patent applications (Baker, 2017a).

What about signal voltages? Here we take the convention that if the pickup is N-up, the sign used for the hum voltage stays the same when the signal voltage is substituted. Conversely, if the pickup is S-up, the sign of the signal voltage substituted for the hum voltage is reversed by −1. So of the pickups between the neck and bridge are N1, S2, S3 and N4, instead of A, B, C and D, for N-up at the neck, S-up in the middle positions and N-up at the bridge, then ignoring the denominators in Voc for c(4), d(4), f(4) and h(4), we have $(A + B − C − D) \rightarrow (n1 − s2 + s3 − n4)$, $(A − B + C − D) \rightarrow (n1 + s2 − s3 − n4)$ and $(A − B − C + D) \rightarrow (n1 + s2 + s3 + n4)$. The first expression can be said to show the difference of two out-of-phase pairs, $(n1 − n4)$ and $−(s2 − s3)$. Or, the sum of the contra-phase pairs, $(n1 − n4)$ and $(s3 − s2)$. The second shows the difference of two in-phase pairs, $(n1 + s2)$ and $−(s3 + n4)$. Or, the sum of two contra-phase pairs, $(n1 − n4)$ and $(s2 − s3)$. It doesn't matter, so long as one recognizes they aren't the same as the first or the third. The third expression shows the sum of two in-phase pairs $(n1 + s2)$ and $(s3 + n4)$.

4.2.2.3 HB Quad Type (2 + 2) Topology Signals

These topologies are less quads than they are two connected pairs, with $J = 4$ number of matched pickups. As such, the calculations are simpler and easier to understand than type (4) topologies. But until someone comes up with a better way to derive and categorize them, we will leave them lumped together with quads.

Circuits e and g each have two pairs of pickups with different coefficients. In circuit e, A and B can be exchanged with each other without changing the tone, but not with C and D. C and D can be exchanged with each other without changing the tone, but not with A and B. The same is true of circuit g. In e and g, the only way to change the tone is to change the sign of one of the pairs or to change the placement of the pickups in the equation. In circuit e, A + B + (C + D)/2 cannot be the same as A + B − (C + D)/2. The same holds true for circuit g. Recall from Sect. 2.5 and Table 2.1 that for J distinct elements, there can be only 2^{J-1} terminal reversals that can produce different tones. Circuits e and g have two separate basic topologies, humbucking pairs, so can have only $2^{2-1} = 2^1 = 2$ different tones from reversing the terminals of the basic topologies.

Now consider A, B, C, and D to be equal hum voltages. In order for Voc = 0, the equations for circuits e and g must have opposite signs for A and B and opposite signs for C and D. So for circuit e, the basic equation for humbucking is Voc = (A − B) ± (C − D)/2 = 0, and the basic equation for humbucking in circuit g is Voc = (A − B)/5 ± 2(C − D)/5 = 0.

What about signal voltages? Here we take the convention that if the pickup is N-up, the sign used for the hum voltage stays the same when the signal voltage is substituted. Conversely, if the pickup is S-up, the sign of the signal voltage substituted for the hum voltage is reversed by −1. So of the pickups between the neck and bridge are N1, S2, S3 and N4, instead of A, B, C and D, for N-up at the neck, S-up in the middle positions and N-up at the bridge, the first signals chosen are (n1 + s2)/5 ± 2(−s3 − n4)/5.

Then one has to cycle through the combinations with (4 things taken 2 at a time) for the (A − B) term times (2 things taken 2 at a time) for the (C − D) term. This equals ((4 ∗ 3)/(2 ∗ 1)) ∗ ((2 ∗ 1)/(2 ∗ 1)) = 6, which is multiplied by $2^{2-1} = 2$ to account for the terminal reversals in the "±" operator, to produce 12 different tone circuit signals: (1) [(n1 + s2) ± 2(−s3 − n4)]/5; (2) [(n1 + s3) ± 2(−s2 − n4)]/5; (3) [(n1 − n4) ± 2(−s2 − s3)]/5; (4) [(−s2 + s3) ± 2(n1 − n4)]/5; (5) [(−s2 − n4) ± 2(−s3 − n1)]/5; and (6) [(−s3 − n4) ± 2(−s2 − n1)]/5. It's usually easier to look at when many of the "−" signs are reversed, i.e., (6) [(s3 + n4) ∓ 2(s2 + n1)]/5. This emphasizes the fact that two matched pickups with opposite poles up are in-phase humbucking pairs.

Alternatively, if one takes just the integer numerator coefficients and writes out all the combinations for 2(A − B) ± (C − D), one gets:

$2A - 2B + C - D = -(2B - 2A - C + D)$	$2B - 2C + A - D = -(2C - 2B - A + D)$
$2A - 2B - C + D = -(2B - 2A + C - D)$	$2B - 2C - A + D = -(2C - 2B + A - D)$
$2A - 2C + B - D = -(2C - 2A - B + D)$	$2B - 2D + A - C = -(2D - 2B - A + C)$
$2A - 2C - B + D = -(2C - 2A + B - D)$	$2B - 2D - A + C = -(2D - 2B + A - C)$
$2A - 2D + B - C = -(2D - 2A - B + C)$	$2C - 2D + A - B = -(2D - 2C - A + B)$
$2A - 2D - B + C = -(2D - 2A + B - C)$	$2C - 2D - A + B = -(2D - 2C + A - B)$

Half of these 24 topologies that can be written down are inverted duplicate circuits, leaving only 12.

4.2.2.4   HB Quad Loaded Tones

Now we can count up the number of potentially different loaded and no-load tones for quad topologies $(2 + 2)$ and (4). Why "potentially different" or "potentially unique"? Because these developments do not guarantee any separation of tones when the strings are fretted and plucked or strummed. Experiments have shown (Sect. 3.3, Fig. 3.9) that some tones from different tone circuits can be very close together.

By Table 4.2, with c(4) and h(4) being equivalent circuits, there are only three different topologies for loaded tones. Let us call that $N_{(4)LT} = 3$. Loaded or not, there are only 3 sets of terminal reversals that will leave the circuits humbucking for type (4) topologies. Let us call that $N_{4SGN} = 3$. Equation (4.4) shows the result. In a patent application, 15/616,396 (Baker, 2017a), N_{4SGN} was called "3 (with pairs & signs)," which came from eliminating duplicates in tables.

Loaded Tones

$$(4) : N_{4LT} * \binom{4}{4} * N_{4SGN} = 3 * 1 * 3 = 9$$

$$(2 + 2) : 2 * \binom{4}{2} * \binom{2}{2} * N_{SGN\text{-}BT} = 2 * 6 * 1 * 2 = 24$$

$$N_{LT\text{-}HQ} = 9 + 24 = 33$$

$$\quad (4.4)$$

$$K_{C\text{-}LT\text{-}HQ} = \binom{K}{4} * N_{LT\text{-}HQ} = \binom{K}{4} * 33, \quad K \geq 4$$

$$\text{for} \quad K = 4, 5, 6, 7, 8; \quad \binom{K}{4} = 1, 5, 15, 35, 70;$$

$$K_{C\text{-}LT\text{-}HQ} = 33, 165, 495, 1155, 2310$$

In the patent application, 15/616,396 (Baker, 2017a), the number 4 was used for N_{4LT}, and the number $N_{T\text{-}HQ} = 36$ was used for the total number of tone circuits, which is now found to be an overestimate of the number of potentially unique tones.

For the $(2 + 2)$ circuits, we have two versions, e$(2 + 2)$ and g$(2 + 2)$ in Figs. 4.13 and 4.14. Given four different pickups, the first pair can be chosen as (4 things taken

2 at a time), or $4 * 3/(2 * 1) = 6$, leaving only one choice, (2 things taken 2 at a time), for the remaining pair. $N_{\text{SGN-BT}}$ is the number of possible terminal reversals of BT number of basic topologies, equal to $2^{\text{BT}-1}$; in this case BT $= 2$. Only the end terminals of the basic topologies can be reversed to change tone. The number of terminal reversals of the two type (2) basic topologies is $2^{2-1} = 2$. That means $2 * 6 * 2 = 24$ possible loaded tones, for a total of $N_{\text{LT-HQ}} = 33$ overall, as shown in Eq. (4.4). For K number of matched single-coil pickups, where K is greater than or equal to 4, there are K things taken 4 at a time times the total number of potentially unique tones, or $K_{\text{C-LT-HQ}}$ number of combinations of potentially unique loaded tones for K number of pickups.

Figure 4.15 shows how a type (2 + 2) circuit topology can produce 6 combinations of (4 things taken 2 at a time) times (2 things taken 2 at a time).

Fig. 4.15 Different combinations of 4 pickups in a type (2 + 2) circuit topology. From Fig. 7, Baker (2017a)

$$AB{}^{C}_{D} \quad AC{}^{B}_{D} \quad AD{}^{B}_{C} \quad BC{}^{A}_{D} \quad BD{}^{A}_{C} \quad CD{}^{A}_{B}$$

4.2.2.5   HB Quad No-Load Tone Circuits

Equation (4.5) shows the calculations for no-load tones. For a type (4) basic topology, there is only 1 no-load tone with 3 different amplitudes, and we are not counting changes in amplitude as changes in tone. Let us call that $N_{(4)\text{NL}} = 1$, and use the amplitudes for other decision-making. It may not be immediately obvious, but e (2 + 2) and g(2 + 2) have the same no-load tone. One only need switch the pairs (A, B) and (C, D) to make $2(A + B) + (C + D)$ into $(A + B) + 2(C + D)$. In Table 4.2, two pickups in series and parallel produce the same no-load tone. Circuits e(2 + 2) and g(2 + 2) are simply the same two basic topologies, in series and parallel, a similar situation. So $N_{\text{NL-HQ}} = 15$ instead of 33.

No-Load Tones

$$(4) : N_{(4)\text{NL}} * \binom{4}{4} * N_{\text{4SGN}} = 1 * 1 * 3 = 3$$

$$(2 + 2) : 1 * \binom{4}{2} * \binom{2}{2} * N_{\text{SGN-BT}} = 1 * 6 * 1 * 2 = 12$$

$$N_{\text{NL-HQ}} = 3 + 12 = 15 \tag{4.5}$$

$$K_{\text{C-NL-HQ}} = \binom{K}{4} * N_{\text{NL-HQ}} = \binom{K}{4} * 15, \quad K \geq 4$$

$$\text{for} \quad K = 4, 5, 6, 7, 8; \quad \binom{K}{4} = 1, 5, 15, 35, 70;$$

$$K_{\text{C-NL-HQ}} = 15, 75, 225, 525, 1050$$

4.2.3 Humbucking Hextets

Figures 4.16, 4.17, 4.18, and 4.19 show the humbucking hextets constructed from the unmatched single-coil triples in Fig. 2.7c. As before, the letters A to F can stand in, as needed, for the pickups, their hum signals, or their string signals, but not for hum and string signals at the same time. Here we start with the largest basic topology type in Fig. 2.7c, the type (3) circuit topologies, and then consider the type $(2 + 1)$ circuit topologies. Note that there are not any obvious duplicates in topologies among the hextets. That remains to be proven by the circuit equations, solved for output, Vo, and the lumped circuit impedance, Zc. For the circuit equations, without explicitly drawing them we will assume that the negative end of the string, according to the "+" signs, is ground and the positive end is Vo, and that there is a load impedance or resistance, R_L, between Vo and ground. Interior voltages nodes, such as V1 and V2, will be added to the equations as necessary. It saves space in the figures.

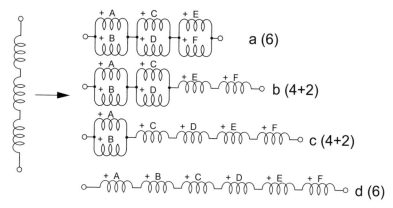

Fig. 4.16 Humbucking hextets a-d, constructed from triple a(3) in Fig. 2.7c

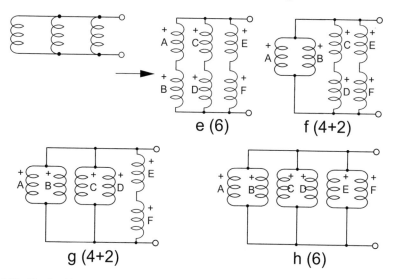

Fig. 4.17 Humbucking hextets e-h, constructed from triple d(3) in Fig. 2.7c

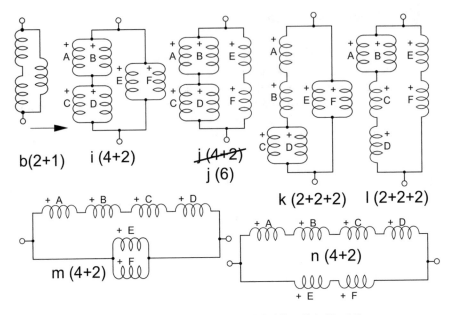

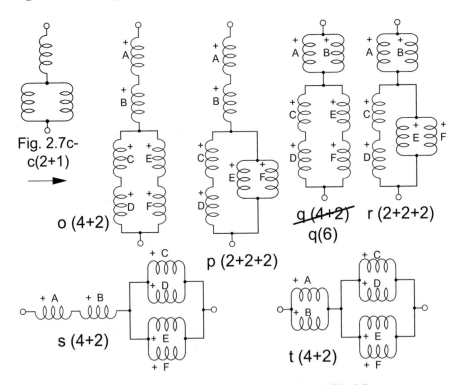

Fig. 4.18 Humbucking hextets i-n, constructed from triple b(2 + 1) in Fig. 2.7c

Fig. 4.19 Humbucking hextets o-t, constructed from triple c(2 + 1) in Fig. 2.7c

Listing 4.6 shows the Maple V entries (without results) for calculations of all the output equations and lumped impedances of the circuits above. The reader should review them and check the math. One can always make mistakes. The results for open-circuit voltage, Voc, the ratio of lumped circuit impedance to pickup impedance, Zc/Z, the Relative intensity and Equivalent circuit (if any) are shown in Table 4.3.

Listing 4.6
Maple V entries for solving the circuit equations for Figs. 4.16–4.19, without results. Assumes that Vo (vo in entries) sits at the "+" output end, with the other end ground, and that a load resistor, R_L (rl in the entries), is connected between Vo and ground. Intermediate node voltages V1 and V2 are added as necessary. The entry voo, or Voo, is the open-circuit voltage which exists when R_L is removed in the limit of $R_L = \infty$. The last line solves for the lumped impedance, Zc, of the pickup circuit, taken to be when the value of R_L loads and drops the open-circuit voltage by half. The reader should review and check the math

```
> restart; #a(6)
> e1 := vo/rl + (vo-a-v1)/z + (vo-b-v1)/z = 0;
> e2 := (v1+a-vo)/z + (v1+b-vo)/z + (v1-c-v2)/z + (v1-d-v2)/z =0;
> solve({e1,e2,e3},{v1,v2,vo});
> vo := rl*(a+b+c+d+e+f)/(3*z+2*rl);
> voo := limit(vo,rl=infinity);
> solve(vo=voo/2,rl);

> restart; #b(4+2)
> e1 := vo/rl + (vo-a-v1)/z + (vo-b-v1)/z = 0;
> e2 := (v1+a-vo)/z + (v1+b-vo)/z + (v1-c-v2)/z + (v1-d-v2)/z =0;
> e3 := (v2+c-v1)/z + (v2+d-v1)/z + (v2-e-f)/2/z = 0;
> solve({e1,e2,e3},{v1,v2,vo});
> vo := 1/2*rl*(a+b+c+d+2*e+2*f)/(3*z+rl);
> voo := limit(vo,rl=infinity);
> solve(vo=voo/2,rl);

> restart; #c(4+2)
> e1 := vo/rl + (vo-a-v1)/z + (vo-b-v1)/z = 0;
> e2 := (v1+a-vo)/z + (v1+b-vo)/z + (v1-c-d-e-f)/4/z=0;
> solve({e1,e2},{v1,vo});
> vo := rl*(a+b+2*c+2*d+2*e+2*f)/(9*z+2*rl);
> voo := limit(vo,rl=infinity);
> solve(vo=voo/2,rl);

> restart; # e(6)
> e1 := vo/rl + (vo-a-b)/2/z + (vo-c-d)/2/z + (vo-e-f)/2/z = 0;
> solve(e1,vo);
> vo := solve(e1,vo);
> voo := limit(vo,rl=infinity);
> solve(vo=voo/2,rl);
```

```
> restart; #f(4+2)
> e1 := vo/rl + (vo-a)/z + (vo-b)/z + (vo-c-d)/2/z + (vo-e-f)/2/z = 0;
> vo := solve(e1,vo);
> voo := limit(vo,rl=infinity);
> solve(vo=voo/2,rl);
> restart; #g(4+2)
> e1 := vo/rl + (vo-a)/z + (vo-b)/z + (vo-c)/z + (vo-d)/z + (vo-e-f)/2/z = 0;
> vo := solve(e1,vo);
> voo := limit(vo,rl=infinity);
> solve(vo=voo/2,rl);

> restart; #h(6)
> e1 := vo/rl + (vo-a)/z + (vo-b)/z + (vo-c)/z + (vo-d)/z + (vo-e)/z + (vo-f)/z = 0;
> vo := solve(e1,vo);
> voo := limit(vo,rl=infinity);
> solve(vo=voo/2,rl);

> restart; #i(4+2)
> e1 := vo/rl + (vo-a-v1)/z + (vo-b-v1)/z + (vo-e)/z + (vo-f)/z = 0;
> e2 := (v1+a-vo)/z + (v1+b-vo)/z + (v1-c)/z + (v1-d)/z = 0;
> solve({e1,e2},{v1,vo});
> vo := 1/2*rl*(a+b+c+d+2*e+2*f)/(z+3*rl);
> voo := limit(vo,rl=infinity);
> solve(vo=voo/2,rl);

> restart; #j(4+2)
> e1 := vo/rl + (vo-a-v1)/z + (vo-b-v1)/z + (vo-e-f)/2/z = 0;
> e2 := (v1+a-vo)/z + (v1+b-vo)/z + (v1-c)/z + (v1-d)/z = 0;
> solve({e1,e2},{v1,vo});
> vo := rl*(a+b+c+d+e+f)/(2*z+3*rl);
> voo := limit(vo,rl=infinity);
> solve(vo=voo/2,rl);

> restart; #k(2+2+2)
> e1 := vo/rl + (vo-a-b-v1)/2/z + (vo-e)/z + (vo-f)/z = 0;
> e2 := (v1+a+b-vo)/2/z + (v1-c)/z + (v1-d)/z = 0;
> solve({e1,e2},{v1,vo});
> vo := rl*(2*a+2*b+c+d+5*e+5*f)/(5*z+12*rl);
> voo := limit(vo,rl=infinity);
> solve(vo=voo/2,rl);

> restart; #l(2+2+2)
> e1 := vo/rl + (vo-a-v1)/z + (vo-b-v1)/z + (vo-e-f)/2/z = 0;
> e2 := (v1+a-vo)/z + (v1+b-vo)/z + (v1-c-d)/2/z = 0;
> solve({e1,e2},{v1,vo});
> vo := rl*(2*a+2*b+4*c+4*d+5*e+5*f)/(10*z+9*rl);
> voo := limit(vo,rl=infinity);
> solve(vo=voo/2,rl);
```

```
> restart; #m(4+2)
> e1 := vo/rl + (vo-a-b-c-d)/4/z + (vo-e)/z + (vo-f)/z =0;
> vo := solve(e1,vo);
> voo := limit(vo,rl=infinity);
> solve(vo=voo/2,rl);
> restart; #n(4+2)
> e1 := vo/rl + (vo-a-b-c-d)/4/z + (vo-e-f)/2/z =0;
> vo := solve(e1,vo);
> voo := limit(vo,rl=infinity);
> solve(vo=voo/2,rl);

> restart; #o(4+2)
> e1 := vo/rl + (vo-a-b-v1)/2/z =0;
> e2 := (v1+a+b-vo)/2/z + (v1-c-d)/2/z + (v1-e-f)/2/z = 0;
> solve({e1,e2},{v1,vo});
> vo := 1/2*rl*(2*a+2*b+c+d+e+f)/(3*z+rl);
> voo := limit(vo,rl=infinity);
> solve(vo=voo/2,rl);

> restart; #p(2+2+2)
> e1 := vo/rl + (vo-a-b-v1)/2/z =0;
> e2 := (v1+a+b-vo)/2/z + (v1-c-d)/2/z + (v1-e)/z + (v1-f)/z = 0;
> solve({e1,e2},{v1,vo});
> vo := rl*(5*a+5*b+c+d+2*e+2*f)/(12*z+5*rl);
> voo := limit(vo,rl=infinity);
> solve(vo=voo/2,rl);

> restart; #q(4+2)
> e1 := vo/rl + (vo-a-v1)/z + (vo-b-v1)/z =0;
> e2 := (v1+a-vo)/z + (v1+b-vo)/z + (v1-c-d)/2/z + (v1-e-f)/2/z = 0;
> solve({e1,e2},{v1,vo});
> vo := rl*(a+b+c+d+e+f)/(3*z+2*rl);
> voo := limit(vo,rl=infinity);
> solve(vo=voo/2,rl);

> restart; #r(2+2+2)
> e1 := vo/rl + (vo-a-v1)/z + (vo-b-v1)/z =0;
> e2 := (v1+a-vo)/z + (v1+b-vo)/z + (v1-c-d)/2/z + (v1-e)/z + (v1-f)/z = 0;
> solve({e1,e2},{v1,vo});
> vo := rl*(5*a+5*b+2*c+2*d+4*e+4*f)/(9*z+10*rl);
> voo := limit(vo,rl=infinity);
> solve(vo=voo/2,rl);

> restart; #s(4+2)
> e1 := vo/rl + (vo-a-b-v1)/2/z =0;
> e2 := (v1+a+b-vo)/2/z + (v1-c)/z + (v1-d)/z + (v1-e)/z + (v1-f)/z = 0;
```

```
> solve({e1,e2},{v1,vo});
> voo := limit(vo,rl=infinity);
> solve(vo=voo/2,rl);

> restart; #t(4+2)
> e1 := vo/rl + (vo-a-v1)/z + (vo-b-v1)/z =0;
> e2 := (v1+a-vo)/z + (v1+b-vo)/z + (v1-c)/z + (v1-d)/z + (v1-e)/z + (v1-f)/z = 0;
> solve({e1,e2},{v1,vo});
> vo := rl*(2*a+2*b+c+d+e+f)/(3*z+4*rl);
> voo := limit(vo,rl=infinity);
> solve(vo=voo/2,rl);
```

Table 4.3 Results of the circuit equation solutions in Listing 4.6, representing Figs. 4.16–4.19

Circuit topology	Solution for Vo	Zc/Z	Relative intensity	Equivalent circuit
a(6)	Vo = (A + B + C + D + E + F)/2	3/2	3	q(6)
b(4 + 2)	Vo = (A + B + C + D + 2(E + F))/2	3	4	o(4 + 2)
c(4 + 2)	Vo = (A + B + 2(C + D + E + F))/2	9/2	5	
d(6)	Vo = A + B + C + D + E + F	6	6	
e(6)	Vo = (A + B + C + D + 2(E + F))/3	2/3	2	j(6)
f(4 + 2)	Vo = (2(A + B) + C + D + E + F)/6	1/3	4/3	i(4 + 2)
g(4 + 2)	Vo = (2(A + B + C + D) + E + F)/9	2/9	10/9	
h(6)	Vo = (A + B + C + D + E + F)/6	1/6	1	
i(4 + 2)	Vo = (A + B + C + D + 2(E + F))/6	1/3	4/3	f(4 + 2)
j(6)	Vo = (A + B + C + D + E + F)/3	2/3	2	e(6)
k(2 + 2 + 2)	Vo = (2(A + B) + C + D + 5(E + F))/12	5/12	4/3	
l(2 + 2 + 2)	Vo = (2(A + B) + 4(C + D) + 5(E + F))/9	10/9	22/9	
m(4 + 2)	Vo = (A + B + C + D + 4(E + F))/9	4/9	12/9	
n(4 + 2)	Vo = (A + B + C + D + 2(E + F))/3	4/3	8/3	
o(4 + 2)	Vo = (2(A + B) + C + D + E + F)/2	3	4	b(4 + 2)
p(2 + 2 + 2)	Vo = (5(A + B) + C + D + 2(E + F))/5	12/5	16/5	
q(6)	Vo = (A + B + C + D + E + F)/2	3/2	3	a(6)
r(2 + 2 + 2)	Vo = (5(A + B) + 2(C + D) + 4(E + F))/10	9/10	22/10	
s(4 + 2)	Vo = (4(A + B) + C + D + E + F)/4	9/4	3	
t(4 + 2)	Vo = (2(A + B) + C + D + D + E)/4	3/4	2	

Note that of the 20 potentially unique topologies, a, b, e, and f have duplicate circuits, leaving 16 unique topologies with 16 potentially unique loaded tones. Relative intensity is calculated by substituting 1 for A to F, each

Table 4.4 Analysis of results in Table 4.3, for topologies in Figs. 4.16–4.19

Topology	Multiplier	(6) 6	(4 + 2) 4	(4 + 2) 2	(2 + 2 + 2) 2	(2 + 2 + 2) 2	(2 + 2 + 2) 2	Zc	Amp
a	1/2	1						3/2	3
d	1	1						6	6
e	1/3	1						2/3	2
h	1/6	1						1/6	1
j*	1/3	1						2/3	2
q*	1/2	1						3/2	3
b	1/2		1	2				3	4
f	1/6		1	2				1/3	4/3
i*	1/6		1	2				1/3	4/3
n	1/3		1	2				4/3	8/3
o*	1/2		1	2				3	4
t	1/4		1	2				3/4	2
m	1/9		1	4				4/9	12/9
s	1/4		1	4				9/4	3
c	1/2		2	1				9/2	5
g	1/9		2	1				2/9	10/9
k	1/12				1	2	5	5/12	4/3
p	1/5				1	2	5	12/5	16/5
l	1/9				2	4	5	10/9	22/9
r	1/10				2	4	5	9/10	22/10

"Multiplier" is the denominator fraction which sets the coefficients of the basic topologies to whole numbers (* duplicate circuit)

Table 4.4 reformats and sorts Table 4.3 to gather all of the type (6), (4 + 2), and (2 + 2 + 2) topologies together in terms of the relative contributions of the basic topologies to each output. The type (6) topologies have only one no-load tone, (A + B + C + D + E + F). The type (4 + 2) topologies are ordered not by the choices of which A, B, etc. labels are used in the basic topologies, but by the coefficients of the (4) and (2) topologies. By this, we can see that there are only 3 no-load tones with basic topology coefficients (1, 2), (1, 4), and (2, 1). By the same token, the (2 + 2 + 2) topologies have only 2 no-load tones, with the basic topology coefficients (1, 2, 5) and (2, 4, 5). So out of 16 unique topologies, there are only 6 unique no-load tones. The rest are the same tones at different amplitudes.

We have found that there is only 1 way that a humbucking pair can be humbucking, and 3 ways that a humbucking quad can be humbucking, based on the cancellation of hum by equal numbers of connections one way and the other. For a humbucking hextet (topologies a, d, e, and h), there are 10 ways that the hextet can be humbucking. If one sequence of the pickups, say A, B, C, D, E, and F, are assigned bits in a binary number, where F is the least significant bit, such that "0" means the same terminal direction, and "1" means the reversed terminal direction, then the binary numbers that have the same number of 0s and 1s are: 000111, 001011, 001101, 001110, 010011, 010101, 010110, 011001, 011010, and 011100. Note that we do not count any binary number beginning with 1, such as 111000, because it is merely the complement or reverse of a binary number already chosen, which means that the signal of the collection is merely inverted. We count the inversions of basic topologies as a separate multiplier, as shown below.

It then remains to could up the number of legitimate potentially unique loaded and no-load tones. Equation (4.6) shows the equations for the loaded tones. From Tables 4.3 and 4.4, we have 4 unique type (6) loaded-tone topologies, 8 unique type (4 + 2) topologies and 4 unique type (2 + 2 + 2) topologies.

Loaded Tones

$$(6) : N_{(6)LT} * \left[\binom{6}{6} * N_{6SGN} \right] = 4 * [1 * 10] = 40$$

$$(4+2) : N_{(4+2)LT} * \left[\binom{6}{4} * N_{4SGN} \right] * \binom{2}{2} * N_{SGN\text{-}BT} = 8 * [15 * 3] * 1 * 2 = 720$$

$$(2+2+2) : N_{(2+2+2)LT} * \binom{6}{2} * \binom{4}{2} * \binom{2}{2} * N_{SGN\text{-}BT} = 4 * 15 * 6 * 1 * 4 = 1440$$

$$N_{LT\text{-}HH} = 40 + 720 + 1440 = 2200$$

$$K_{C\text{-}LT\text{-}HH} = \binom{K}{6} * N_{LT\text{-}HH} = \binom{K}{6} * 2200, \ \ K \geq 6$$

$$\text{for } K = 6, 7, 8; \binom{K}{6} = 1, 7, 28;$$

$$K_{C\text{-}LT\text{-}HH} = 2200, \ 15{,}400, \ 61{,}600 \tag{4.6}$$

Equation (4.6) covers loaded tones in humbucking hextets. $N_{6SGN} = 10$ is the number of humbucking hexes that can be wired merely by reversing the terminals of individual pickups within a type (6) to obtain different humbucking circuits, with 3 hum signals of one polarity and 3 of the other. Note that in a type (4 + 2) topology, the same thing occurs for the basic topologies of type (4). $N_{SGN\text{-}BT} = 2^{BT-1}$, where BT is the number of basic topologies in the circuit. Here we get 2200 potentially unique loaded tones.

No-Load Tones

$$(6) : N_{(6)NL} * \left[\binom{6}{6} * N_{6SGN} \right] = 1 * [1 * 10] = 10$$

$$(4+2) : N_{(4+2)NL} * \left[\binom{6}{4} * N_{4SGN} \right] * \binom{2}{2} * N_{SGN\text{-}BT} = 3 * [15 * 3] * 1 * 2 = 270$$

$$(2+2+2) : N_{(2+2+2)NL} * \binom{6}{2} * \binom{4}{2} * \binom{2}{2} * N_{SGN\text{-}BT} = 2 * 15 * 6 * 1 * 4 = 720$$

$$N_{NL\text{-}HH} = 10 + 270 + 720 = 1000$$

$$K_{C\text{-}NL\text{-}HH} = \binom{K}{6} * N_{NL\text{-}HH} = \binom{K}{6} * 1000, \ \ K \geq 6$$

$$\text{for } K = 6, 7, 8; \ \ \binom{K}{6} = 1, 7, 28;$$

$$K_{C\text{-}NL\text{-}HH} = 1000, \ 7000, \ 28{,}000 \tag{4.7}$$

Equation (4.7) covers no-load tones in humbucking hextets. The changes consist entirely of the number of no-load tones, 1 for (6), 3 for (4 + 2) and 2 for (2 + 2 + 2), leaving less than half the number of loaded tones, 1000. Recall that the same no-load tones come still from circuits with different lumped impedances and relative signal amplitudes. This leaves an engineering choice as to which circuits will be picked to avoid wide ranges in output amplitude or active amplitude compensation.

4.2.4 Humbucking Octets

Humbucking Octets and up are left mostly as an exercise for the reader; you now have the tools to construct and analyze them. But here is a table from the second of the patent applications counting the subcategories of topologies, with their constructions of basic topologies, as derived from converting the unmatched single-coil topologies in Fig. 2.8a, b–i to j into matched pairs.

These numbers should not be taken for granted and should be checked, for two reasons: (1) circuit equations and solutions, like those in Listing 4.6, were not run on these constructions, and the placement of various circuit topologies in different

Table 4.5 Compilation of humbucking octet topologies from a patent application, constructed from applying Fig. 4.11 to 2.8a, b, topologies i to j, with the number and sums of resulting constructions of basic topologies from (8) to (2 + 2 + 2 + 2)

	a	b	c	d	e	f	g	h	i	j	Sums
(8)	2							2	2	2	8
(6 + 2)	2	4					4	2			12
(4 + 4)	1			4	4			1		1	11
(4 + 2 + 2)		4	8	4	4	8	4		2	2	36
(2 + 2 + 2 + 2)			4	1	1	4				1	11
Sums	5	8	12	9	9	12	8	5	5	5	78

From Math 27, Baker (2017a)

subcategories can be mistaken; (2) anyone can make mistakes. Consider the duplicate, equivalent circuits in quads and hextets. In the quads, Fig. 4.14c, h turned out to be equivalent. The only difference between them is a connection between the middle taps of the two coils in 4.14h on the left and right. Note that both topologies, which look like types (2 + 2), are symmetrical vertically and horizontally. So the coils could be "moved" around in part merely by flipping the topology horizontally and vertically. And the circuit equations confirm that they are indeed type (4) topologies.

In the hextets, the equivalent circuits Figs. 4.16-a and 4.19-q both have symmetrical quads. As do 4.16-b and 4.19-o, 4.16-f and 4.17-i, and 4.17-e and 4.18-j. The original construction of Table 4.5 attempted to take account of such symmetries, but they were not confirmed by solutions of circuit equations. In the original patent work, the cumbersome table method (not presented here) calculated the number of humbucking circuits in a type (8) topology by the terminal reversals of sets of 4 coils as 35. This has been confirmed by counting the number of 8-bit binary numbers below 10000000, with an equal number each of four 0s and 1s.

In a visual inspection of notes on humbucking octets, from which Table 4.5 comes, all topologies in which a circuit like Fig. 4.14c(4) were counted out as having likely duplicate circuits. This left sums for the right-hand column of Table 4.5 as: (8) 6, (6 + 2) 9, (4 + 4) 4, (4 + 2 + 2) 27, and (2 + 2 + 2 + 2) 11. Using these numbers, the number of loaded tones are estimated in Eq. (4.8), but still require confirmation.

Loaded Tones estimate

$$(8): N_{(8)LT} * \left[\binom{8}{8} * N_{8SGN} \right] = 6 * [1 * 35] = 210$$

$$(6 + 2): N_{(6+2)LT} * \left[\binom{8}{6} * N_{6SGN} \right] * \binom{2}{2} * N_{SGN\text{-}BT}$$

$$= 9 * [28 * 10] * 1 * 2 = 5040$$

$$(4 + 4): N_{(4+4)LT} * \left[\binom{8}{4} * N_{4SGN} \right] * \left[\binom{4}{4} * N_{4SGN} \right] * N_{SGN\text{-}BT}$$

$$= 4 * [70 * 3] * [1 * 3] * 2 = 5040$$

$$(4 + 2 + 2): N_{(4+2+2)LT} * \left[\binom{8}{4} * N_{4SGN} \right] * \binom{4}{2} * \binom{2}{2} * N_{SGN\text{-}BT}$$

$$= 27 * [70 * 3] * 6 * 4 = 136{,}080$$

$$(2 + 2 + 2 + 2): N_{(2+2+2+2)LT} * \binom{8}{2} * \binom{6}{2} * \binom{4}{2} * \binom{2}{2} * N_{SGN\text{-}BT}$$

$$= 11 * 28 * 15 * 6 * 8 = 221{,}760$$

$$N_{LT\text{-}HO} = 210 + 5040 + 5040 + 136{,}080 + 221{,}760 = 368{,}130$$

$$K_{C\text{-}LT\text{-}HO} = \binom{K}{8} * N_{LT\text{-}HO} = \binom{K}{8} * 368{,}130, \quad K \geq 8$$

$$\text{for} \quad K = 8, 9, 10; \quad \binom{K}{8} = 1, 9, 45;$$

$$K_{C\text{-}LT\text{-}HO} = 368{,}130, \quad 3{,}313{,}170, \quad 16{,}565{,}850 \tag{4.8}$$

No-load tones are left as an exercise for the reader.

4.2.5 Compilation of Results

In the matter of numbers of humbucking circuits with pickups terminal reversals for topologies of type (J), where J is even and $J > 2$, i.e., $N_{4SGN} = 3$, $N_{6SGN} = 10$, and $N_{8SGN} = 35$, the previous patent work found that the number could be expressed by a combination of $(J - 1$ things taken $(J/2 - 1)$ at a time), as shown in Eq. (4.9), and expressed as N_{JSGN} in Eqs. (4.4)–(4.7). It was extended to $J = 10$ and higher by induction, and is presented here as a conjecture, without mathematical proof. Note that the numbers tend to go up exponentially, as do many results here.

$$(J) = (4), (6), (8), (10), \ldots, J \text{ even}$$

$$N_{JSGN} = \binom{3}{1}, \binom{5}{2}, \binom{7}{3}, \binom{9}{4}, \ldots, \binom{Je-1}{\dfrac{Je}{2}-1} \tag{4.9}$$

$$= 3, \ 10, \ 35, \ 126, \ldots$$

We can compile results of loaded tones, using the estimates for humbucking octets, but for no-load tones, there is not enough information yet available to estimate anything for octets, without solving all the 78 sets of circuit equations. Table 4.6 shows the results for humbucking loaded tones, up to hextets, and Table 4.7 shows the results for humbucking no-load tones, up to hextets.

Table 4.6 Numbers and sums for loaded tones for humbucking pairs, quads, hextets, and octets, for numbers of matched single-coil pickups from 2 to 8

$J =$	2	4	6	
$N_{LT} =$	2	33	2200	
K	Pairs	Quads	Hextets	Sums
2	2			2
3	6			6
4	12	33		45
5	20	165		185
6	30	495	2200	2725
7	42	1155	15,400	16,597
8	56	2310	61,600	63,966

The inner numbers are N_{LT} times (K pickups taken J at a time). The sums show how many potentially unique loaded tones can be had K pickups using pairs, quads and hextets. Adapted and corrected from Math 31, Baker (2017a)

Table 4.7 Numbers and sums for no-load tones for humbucking pairs, quads, and hextets, for numbers of matched single-coil pickups from 2 to 8

$J =$	2	4	6	
$N_{NL} =$	1	15	1000	
K	Pairs	Quads	Hextets	Sums
2	1			1
3	3			3
4	6	15		21
5	10	75		85
6	15	225	1000	1240
7	21	525	7000	7546
8	28	1050	28,000	29,078

Recall that all these humbucking tones are "potentially unique" tones.

References

Baker, D. L. (2016b). Acoustic-electric stringed instrument with improved body, electric pickup placement, pickup switching and electronic circuit, US Patent 9,401,134, 26 July 2016. Retrieved from https://patents.google.com/patent/US9401134B2/

Baker, D. L. (2017a). Humbucking switching arrangements and methods for stringed instrument pickups, US Patent Application 15/616,396, filed 7 June 2017, published as US-2018-0357993-A1, 13 Dec 2018, granted as Patent US10,217,450, 26 Feb 2019. Retrieved from https://www.researchgate.net/publication/335727402_Humbucking_switching_arrangements_and_methods_for_stringed_instrument_pickups_-_NPPA_15616396

Baker, D. L. (2018b). Means and methods for switching odd and even numbers of matched pickups to produce all humbucking tones, US Patent Application 16/139,027, 22 Sep 2018, published as US-2019-0057678-A1, Feb 21, 2019, granted as U.S. Patent 10,380,986, 08/13/2019. Retrieved from https://www.researchgate.net/publication/335728060_NPPA-16-139027-odd-even-HB-pu-ckts-2018-06-22

Baker, D. L. (2018c). Means and methods for obtaining humbucking tones with variable gains, US Patent Application 16/156,509, 10 Oct 2018, published as US-2019-0057679-A1, 21 Feb 2019, Pending. Retrieved from https://www.researchgate.net/publication/333520942_Means_and_methods_for_obtaining_humbucking_tones_with_variable_gains_Non-Provisional_Patent_Application_16156509_of_Donald_L_Baker_Tulsa_OK

Fender, C. L. (1966). Electric guitar incorporating improved electromagnetic pickup assembly, and improved circuit means, US Patent 3,290,424, 6 Dec 1966. Retrieved from https://patents.google.com/patent/US3290424A/

Chapter 5
The Limits of Mechanical Switches

5.1 Some History

The References in Chapter 2 contain patent references to a number of guitar switching systems. One did ask a USPTO Patent Examiner if the Jacob application (US-2009-0308233-A1) interfered in any way with some of the following circuits, but never got an answer. Instead, the Examiner touted Wnorowski (US6,998,529 2006) as likely interfering, but could offer no substantive reasons as to why.

Figures 4.1–4.3 show one way to concatenate switches to increase the number of possible outputs from 5 to10. But it fills the available real estate under an electric guitar pickguard. In most electric guitars, the controls have to be placed both out of

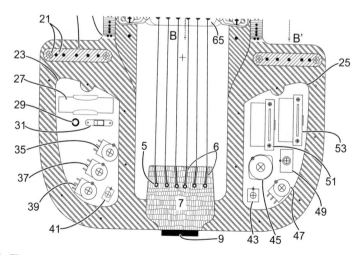

Fig. 5.1 Electronics cavities and design placement of battery (27) and electronic controls, with two 5-way superswitches (51 and 53) and a 4P2T switch to choose between them (49). From Fig. 1 of Baker (2016b)

© Springer Nature Switzerland AG 2020
D. L. Baker, *Sensor Circuits and Switching for Stringed Instruments*,
https://doi.org/10.1007/978-3-030-23124-8_5

the way of the picking area, and still convenient to reach nearby. The somewhat ugly prototype guitar for US Patent 9,401,134 (Baker, 2016b) accomplished this by making the butt end of the guitar large and thick, with two thin steel plates mounted on either side of the skinny end of a paddle-shaped sound board (Figs. 5.1, 5.2, 5.3, and 5.4). Workable, but voted least likely to be seen at a NAMM show.

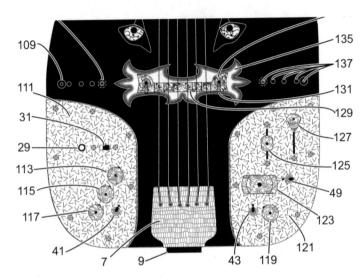

Fig. 5.2 Design placement electronic controls on two steel plates (111 and 121) in US Patent 9,401,134. From Fig. 6 in Baker (2016b)

Fig. 5.3 Actual placement electronic controls on two steel plates, including the 12-way tone capacitor and clipping diode knob, battery light, battery switch, two 5-way superswitches, 4P2T selector switch for the 5-way switches, the linear-distortion circuit switch and two pots for volume and distortion control. The back plate and ¼-in. phono output jack can just be seen to the left

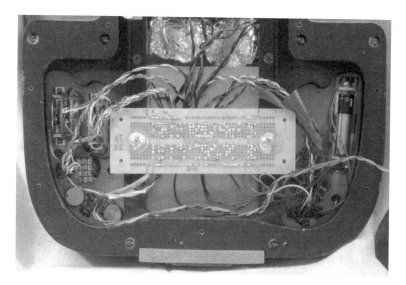

Fig. 5.4 Printed circuit board and wiring rat's nest with back cover removed

5.2   Concatenating Switches

Figures 3.7 and 3.8 show concatenated switches for a dual-humbucker, 24-way switching system with 20 different series-parallel combinations of the four humbucker coils. In Fig. 3.7, a 4P2T switch on each humbucker switches between series and parallel combinations of the two humbucker coils, with optional peaking capacitors. In Fig. 3.8, a 4P6T rotary switch takes the two 2-wire outputs of the 4P2T switch on each humbucker, and produces two outputs (in-phase and contra-phase) with the humbuckers in series, two outputs (in-phase and contra-phase) with the humbuckers in parallel, and two outputs with one humbucker each.

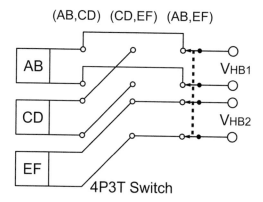

Fig. 5.5 A 4-pole, 3-throw rotary switch selects 2 of 3 humbuckers, each with the series-parallel switch from Fig. 3.7, feeding V_{HB1} and V_{HB2} into the humbucker AB and CD inputs in Fig. 3.8. From Fig. 16 of Baker (2017a)

It is possible to add a third humbucker with the circuit in Fig. 5.5, using a 4 pole, 3-throw rotary switch. This means 3 combinations of 2 humbuckers each, with 4 series-parallel combinations of the internal humbucker coils, times 4 series-parallel combinations of 2 humbuckers, for 48 combinations of 4 of the six humbucker coils. Plus, there are 2 series-parallel combinations of each of the 3 humbuckers alone for another 6 circuits, a total of 54 unique circuit combinations of six coils out of a total of 144 possible switch configurations.

The 144 switch configurations comes from counting all those that can be thrown, even when the pickup attached to the switch is not in the circuit. Three humbuckers, each with a 2-throw switch (Fig. 3.7), can be configured $2^3 = 8$ ways. The switch in Fig. 5.5 can be configured 3 ways. The switch in Fig. 3.8 can be configured 6 ways. So $8 * 3 * 6 = 144$, from three 4P2T toggle switches, a 4P3T rotary switch and a 4P6T rotary switch. Then measure and count the number of potentially unique tones that are too close together to count as separate and subtract them from 54.

This is one of the limitations of designs with electromechanical switches. The more coils switched and the more switches used, the more duplicate outputs produced. It gets even worse if the circuit designer did not pay attention to the circuits and added duplicate circuits to the design, which can be compounded if the switches feed into active amplifiers with high impedances, which tend to push the roll-off frequency differences between series and parallel circuits above human hearing.

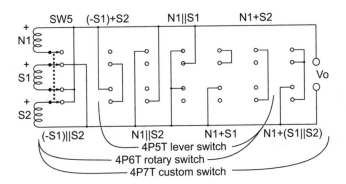

Fig. 5.6 Seven possible series-parallel humbucking circuits for 3 matched single-coil pickups, with choices for existing 4P5T and 4P6T switches, with a possible 4P7T concatenated switch, Fig. 5.7. From Fig. 18 of Baker (2017a)

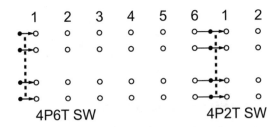

Fig. 5.7 Making a 4P7T switch by concatenating a 4P6T rotary switch with a 4P2T toggle switch. From Fig. 19 of Baker (2017a)

Figure 5.6 shows seven possible series-parallel humbucking circuits for 3 matched single-coil pickups, and choices for 4-pole, 5-, 6-, and 7-throw switches. The 5-pole switch exists as a 5-way superswitch, available from guitar parts companies allparts. com and stewmac.com. The 6-pole switch exists as a 4P6T rotary switch with 2 wafers. It takes up even more room under a pickguard than the 5-way switch. A 7-pole switch can be constructed by letting the last throw of a 4P6T rotary switch connect to the poles of a 4P2T toggle switch, which then has 2 choices to add to the 5 on the rotary switch.

It is a reasonable design choice in that the switches can fit under a pickguard, even if the electronics hole has to be enlarged. But, since the actual tones which result will tend to bunch at the warm end, it does not guarantee that all seven choices are justified.

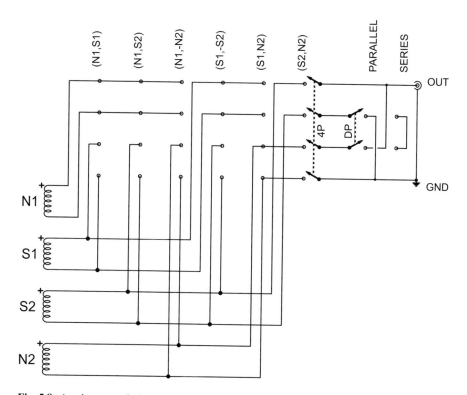

Fig. 5.8 An alternate solution to Figs. 4.1–4.3, using a 4-pole, 6-throw rotary switch to produce humbucking pairs of 4 matched coils, feeding into a 4-pole, double-throw toggle switch to produce series and parallel circuits of the two chosen coils. From Fig. 8 of Baker (2016a)

Figure 5.8 shows an alternate solution to the two 4P5T superswitches, with a 4P2T selector toggle switch, used in Figs. 4.1–4.3. It uses a 4P6T rotary switch in the manner of Fig. 5.5 to choose the six possible combinations of four matched single-coil pickups, and connects them to the throws of the rotary switch so that the 4P2T

toggle switch can connect them together either in series or parallel. Note that when
the two pickups have the same pole up, (N1, −N2) or (S1, −S2), then the second
pickup is reversed to produce humbucking contra-phase signals. This was not
possible with the setup in Figs. 4.1 and 4.2, where the pickups were connected to
the poles, and denied 2 of the 6 choices. So the 5th throws produced in-phase
humbucking quads. Note also that the sequence of throws in Fig. 5.8 has nothing
to do with the warmth of the tones; it merely makes connections in another logical
order.

Figure 5.9 shows a general diagram for concatenating a 4-pole, M1-throw switch
with a 4-pole, M2-throw switch, to get M1 + M2 − 1 throws.

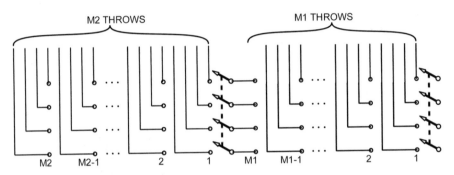

Fig. 5.9 A g general diagram for concatenating a 4-pole, M1-throw switch with a 4-pole,
M2-throw switch, to get M1 + M2 − 1 throws

5.3   General Characteristics of Rotary Switches

Rotary switches typically are made with non-conducting wafers with throw contacts
spaced at 30° intervals, or 12 per 360° rotation. A smaller secondary wafer typically
carries the poles, which rotate passing through the contacts. They tend to be made as
1-pole, 12-throws, or 2-poles, 6-throws, or 3-poles, 4-throws, or 4-poles, 3-throws
per wafer. Thus, a 4P6T rotary switch will have 2 wafers of 2 poles and 6 throws.
The wafers tend to be about an inch (25 mm) in diameter, with solder or printed
circuit terminals adding to the diameter. A standard 5-way guitar switch or 4P5T
superswitch is a rotary switch mounted on its side, with a lever actuator instead of a
shaft. There are a set of miniature enclosed rotary switches that go up to 6P6T,
but the 6P6T version is typically non-stocked and special order, and costs on the
order of $50.

Depending on the circuit construction used, 3 single-coil pickups can have 6–9
different humbucking circuit configurations, and 4 single-coil pickups can have
25–31 different humbucking circuit constructions. We will see later that not all of
the circuits will have significantly different tones, and can limit the useful number of
tones for 4 matched single-coil pickups to about 12–16. Therefore the mechanical

switches that will fit under the typical solid-body electric guitar pickguard cannot accommodate switching much over 3 or 4 coils, with the expectation of getting a full range of humbucking tones.

One very slick-looking guitar on the market comes equipped with controls like a sound engineer's panel. But if one remembers correctly they cover the whole off-hand side of the guitar, below the 1-string, from the bridge to the neck. The body in that area sets at a down-angle so that strumming will not hit the controls. Thus, they are not necessarily as convenient to hand as standard controls.

It is possible to do a great deal with mechanical guitar controls, but above 4 coils, they cannot fulfill the entire range of possibilities in humbucking circuits.

References

Baker, D. L. (2016a). A switching system for paired sensors with differential outputs, especially matched single coil electromagnetic pickups in stringed instruments, US Provisional Patent Application 62/355,852, 28 June 2016. Retrieved from https://www.researchgate.net/publication/335727679_Provisional_Patent_Application_A_Switching_System_for_Paired_Sensors_with_Differential_Outputs_Especially_Matched_Single_Coil_Electromagnetic_Pickups_in_Stringed_Instruments

Baker, D. L. (2016b). Acoustic-electric stringed instrument with improved body, electric pickup placement, pickup switching and electronic circuit, US Patent 9,401,134, 26 July 2016. Retrieved from https://patents.google.com/patent/US9401134B2/

Baker, D. L. (2017a). Humbucking switching arrangements and methods for stringed instrument pickups, US Patent Application 15/616,396, filed 7 June 2017, published as US-2018-0357993-A1, 13 Dec 2018, granted as Patent US10,217,450, 26 Feb 2019. Retrieved from https://www.researchgate.net/publication/335727402_Humbucking_switching_arrangements_and_methods_for_stringed_instrument_pickups_-_NPPA_15616396

Jacob, B. L. (2009). Programmable switch for configuring circuit topologies, US Patent Application 2009/0308233 A1, 17 Dec 2009. Retrieved from https://patents.google.com/patent/US20090308233A1/

Wnorowski, T. F. (2006). Method for switching electric guitar pickups, US Patent 6,998,529 B2, 14 Feb 2006. Retrieved from https://patents.google.com/patent/US6998529B2/

Chapter 6
An Efficient uC-controlled Cross-Point Pickup Switching System

6.1 Some Prior Art

This author knows only of two other programmable processor systems covered by patents with the intent of switching larger numbers pickup circuits on electric guitars than can be managed by electromechanical switches: Bro & Super (US7,276,657, 2007) and Ball et al. (US9,196,235, 2015; 9,640,162, 2017). Both systems claim to use commonly existing controls on electric guitars, which the uC then interprets to its own uses. The Bro and Super system uses micro relays to switch pickup coils. It claims in Table 1 to produce 156 tonal outputs from the four coils of two dual-coil hum buckers, about half of which are humbucking. For 5 coils, it claims 474 tonalities, and 6, it claims 1106 tonalities. At one time, the commercial product could be found at https://guitarelectronics.com, but at the time of this writing has been removed.

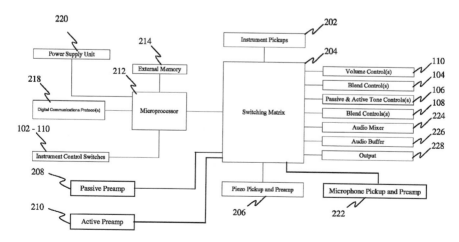

Fig. 6.1 From Fig. 2, Ball et al. (US9,196,235, 2015; 9,640,162, 2017), showing the block diagram for a uC guitar pickup switching and control system

© Springer Nature Switzerland AG 2020
D. L. Baker, *Sensor Circuits and Switching for Stringed Instruments*,
https://doi.org/10.1007/978-3-030-23124-8_6

Ball et al. (US9,196,235, 2015; 9,640,162, 2017) patented a uC system for switching guitar pickups and controlling sound, with 17 Claims and 3 Figures, of the kind of detail shown in Fig. 6.1 above. In various commercial incarnations, it allowed for combinations of two humbuckers, a single-coil pickup and a piezoelectric pickup. See gamechanger.music-man.com/specs.eb. Some advertising claimed "over 250,000 pickup combinations" (https://www.music-man.com/instruments/guitars/the-game-changer). The patent claims "any combination of combinations" of pickups (2015, col. 4) and "any/every possible combination of serial, parallel, in phase, and out of phase" of pickup circuit (2015, col. 7).

This author questions the 250,000 figure, as Table 2.4 here shows the number of series-parallel circuits to be about 23 times lower. To the knowledge of this author, the patent owner has not produced any public white paper or other description identifying the "250,000 pickup combinations," how they are constructed, or any measure of tonality. Nor has the company involved responded definitively to direct query from this author to offer any explanation.

Note that according to Fig. 6.1, most functions and signals are routed through the "digitally controlled analog switching matrix" ((204), col. 4, 2015). The patent is short, 10 pages long, and does not address the system in detail, not explaining, for example, how the "switching matrix" can be used for both control and signal lines without causing dropouts in either one. While the system may be capable of connecting any combination of 4–6 pickups into any useable circuit, the patent owner has apparently elected to keep the methods proprietary.

6.1.1 An Efficient uC-Controlled Cross-Point Pickup Switching System

Figure 6.2 shows the block diagram of a system architecture for achieving the aim of this book—a switching system for any number of pickups or coils, with which the guitarist/pianist/user can shift monotonically from bright to warm tones and back, without ever having to know which pickups are used in what combinations. Note that the pickup switching and control inputs are separated on the uC, and that it does not assume that only familiar guitar controls need be used, as did the Bro & Ball patents. It allows other types of controls and displays, for example, mouse-like wheels and buttons and digital displays, or a touch-activated swipe and tap sensor integrated a color digital display, like that on a smart cell phone.

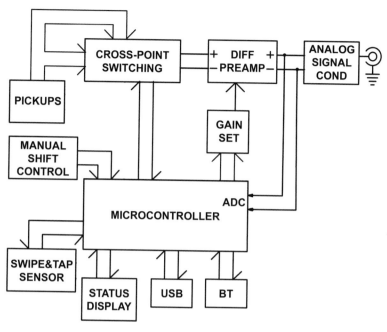

Fig. 6.2 From Fig. 20, US Patent Application 15/616,396 (Baker, 2017a). A solid-state cross-point switch, controlled by a uC, combines both terminals each of a set of pickup coils to produce pickup circuits (preferably humbucking). The two output terminals of the coil circuit are fed into a differential amplifier, with the uC controlling gain to equalize signal strength for the different circuits. The output of the differential amplifier feeds into an analog-to-digital converter (ADC) on the uC for sampling signals to produce FFT spectra, and into the output analog signal conditioning. The uC has a manual tone shift control and/or a touch-activated swipe & tap sensor, which may or may not be covering a digital status display. The uC also has serial input-output ports, such as USB and BlueTooth

A digitally controlled analog cross-point switch with Mx x-inputs and My y-inputs, has Mx times My cross-point interconnections with 2^{Mx*My} switch choices. All of the pickup terminals are connected to both the x and y inputs, with at least two extra for the outputs. So for Mx pickup terminals, My must be greater than or equal to Mx + 2 to account for the two output terminals. For example, if there are 4 humbuckers, or 4 humbucking pairs, then Mx must be at least 8, and My must be at least 10. And the inherent 2^{80} or more interconnections choices is a very large number, well beyond the needs of the pickup switching discussed here. It can accommodate all manner of humbucking and non-humbucking circuits, including those developed here in Chap. 2.

Commonly available cross-point switches, such as the Zarlink MT093 iso-CMOS 8 × 12 analog switch array, ~$7/each, and the Intersil CD22M3494 BiMOS 16 × 8 cross-point switch, ~$6.50/each, require digital sequencing and control for the cross-point switch array. This means a micro-controller, particularly a low-power micro-controller. It is possible to concatenate cross-point switches to form, for example, a

16 × 16 from two 8 × 16 cross-point switches. Subtracting 2 for the output, that leaves 14 pickup terminals to connect, either 7 matched single-coil pickups, or any combination of 7 humbucker and matched single-coil pickups. For 8 single-coil pickups, or 4 humbuckers with all four terminals, plus an output connected to the cross-point switch, a 16 × 18 or larger switch is needed, such as four 8 × 12 switches concatenated into a 16 × 24 switch, or three 8 × 16 switches into a 16 × 24 switch.

Here one parts company with Bro & Super (US7,276,657, 2007), Jacob (US-2009-0308233-A1), and Ball et al. (US9,196,235, 2015; 9,640,162, 2017). It is neither necessary nor desirable either to have separate pickup and circuit selection, as Jacob required, nor to have input and output controls look exactly the same as current guitars and basses and other stringed instruments, as Ball required. A cross-point switch combines both selection and connection circuits. And using only classic controls with knobs can be too limiting, requiring more control surface space than actually needed. Routing controls through the analog switching matrix means either that the number of possible vibration sensors, or pickups, will be limited by and compete with the number of control lines, or that the size of the analog switching matrix must be quite large to handle any number of pickups above 3 or 4, plus 3–5 controls. It is more efficient to use digital controls and multiplexers, connected directly to the micro-controller, which can also provide any drive signals necessary. Modern digital mice and smart phones are a good example.

Figure 6.2 shows a different concept. Instead of nearly every control going through the switching matrix, as in the Ball patents, all go through the microcontroller. Only the pickups, or any other sensors, and the microcontroller provide inputs to the cross-point switch. The box "PICKUPS" refers to any number and kind of sensors. The only output from the cross-point switch is connected to a differential amplifier with a gain set by the microcontroller. The "analog signal conditioning" block can be as simple as a volume pot, or add more complex audio tone and aftereffects circuits. The "manual shift control" is the most basic control. It can be embodied as merely a binary up-down, debounced, momentary contact toggle switch, or push buttons, that triggers a count up and down through a preset sequence of pickup combinations. The most basic output for a "status display" is a set of binary lights, controlled by the micro-controller, which merely turn on or off to indicate the position of the selection in the sequence. It could also be an alphanumeric segment display, or pixel array display, especially if the selection sequence is more than 6 or 8 long.

But much more is possible. The "manual shift control" could be like the scrolling wheel on a mouse, with rotation to change selection and the down, left and right switches to change modes, such as setting presets of favorite tones, and moving tones up and down in the selection sequence. That kind of input could also be done with a "swipe & tap sensor," with a "status display" that shows alphanumeric data to indicate presets and selections, or done with a touch-sensitive screen like a smart phone, built into the stringed instrument. This could also be done with USB or Bluetooth, BT, or other digital connections, which could also be used to diagnose and reprogram the microcontroller, if needed.

Most if not all current microcontrollers have an analog-to-digital converter, or ADC. In US 9,401,134 (Baker, 2016b) pickup position can be changed to any position, attitude, and height between the neck and bridge. This would change any bright-to-warm preset sequence of humbucking or other combinations. So would changing the model of any of the pickups. So the ADC converter is used to perform frequency spectrum analysis on the results, to aid in re-ordering the selection sequence from bright to warm. And if it becomes hopelessly confused, the mode switch setting on the "manual shift control" or the "swipe & tap sensor" can be used to reset the sequence to a factory setting.

Using a fast-Fourier transform, or FFT, computed by the micro-controller, spectral analysis could be done by manual strumming of the stringed instrument, or by means of an automatic strumming device, attached to the stringed instrument and controlled by the micro-controller via USB or another digital connection. As of this writing (November 2018) the use of FFTs to order the tones of different switching combinations, from bright to warm, is still a matter of research. Equation 3.4 in Sect. 3.3.3 shows a preliminary approach which must be improved with existing psychoacoustic research, which could be modified by the musician with presets or re-ordering of the sequence, should perception prove different. This will also help identify which tonal outputs may be duplicates, and thus may be excluded from the sequence.

References

Baker, D. L. (2016b). Acoustic-electric stringed instrument with improved body, electric pickup placement, pickup switching and electronic circuit, US Patent 9,401,134, 26 July 2016. Retrieved from https://patents.google.com/patent/US9401134B2/

Baker, D. L. (2017a). Humbucking switching arrangements and methods for stringed instrument pickups, US Patent Application 15/616,396, filed 7 June 2017, published as US-2018-0357993-A1, 13 Dec 2018, granted as Patent US10,217,450, 26 Feb 2019. Retrieved from https://www.researchgate.net/publication/335727402_Humbucking_switching_arrangements_and_methods_for_stringed_instrument_pickups_-_NPPA_15616396

Ball, S., et al. (2015). Musical instrument switching system, US Patent 9,196,235 B2, 24 Nov 2015. Retrieved from https://patents.google.com/patent/US9196235B2/

Ball, S., et al. (2017). Musical instrument switching system, US Patent 9,640,162 B2, 2 May 2017. Retrieved from https://patents.google.com/patent/US9640162B2/

Bro, W. J., & Super, R. L. (2007). Maximized sound pickup switching apparatus for string instrument having a plurality of sound pickups. US Patent 7,276,657 B2, October 2, 2007.

Jacob, B. L. (2009). Programmable switch for configuring circuit topologies, US Patent Application 2009/0308233 A1, 17 Dec 2009. Retrieved from https://patents.google.com/patent/US20090308233A1/

Chapter 7
The Tonal Advantages of Pickups with Reversible Magnets

7.1 Introduction and Some Prior Art

By rights, this material should be more than one chapter, but the creative and inventive process sometimes gets tangled and has to be fixed. It comes initially from the Non-Provisional Patent Application 15/917,389 (Baker, 2018a), which covered an invention of matched single-coil pickups with reversible magnets, and the construction of circuits to generate 2^{J-1} number of different tonal characters from J number of pickups.

For example, if one has 3 matched single-coil pickups with reversible magnets, with all north poles up (N-up), labeled N1, N2, and N3 from bridge to neck, then one can get 3 humbucking pairs and 3 humbucking triples, all primarily contra-phase or out-of-phase. One can get only 3 more sets: S1, N2, N3; N1, S2, N3; and N1, N2, S3, where S1, S2, and S3 represent pickups with S-up poles in those positions. All the other 4 combinations of N-up and S-up pickups duplicate the tonal characters of those already constructed, because the outputs are merely reversed in phase. And if two pickup sets share the same pickups, for example, N1 and N3, then those set share some of the same tones.

Later, in developing Chap. 4, this author determined a new way to look at pickup circuits, using open-circuit output voltages to determine no-load tones, and advanced that development in writing this chapter. Now the large numbers of potentially unique humbucking and non-humbucking series-parallel circuits can be reduced to just a few no-load circuit equations to express tone, modified by the position placements of pickups in those equations. This first part of this chapter covers the rules for humbucking expressions using open-circuit output equations and relations.

The chapter then covers the development of humbucking pairs, triples, quads, quintets (quints), and hextets (hexes) using open-circuit output relations, followed by the development and application of magnet reversals to those equations, and the description of two embodiments of matched single-coil pickups with reversible magnets.

© Springer Nature Switzerland AG 2020
D. L. Baker, *Sensor Circuits and Switching for Stringed Instruments*,
https://doi.org/10.1007/978-3-030-23124-8_7

In searching patents, about 9 months after the patent application 15/917,389 (Baker, 2018a) was filed, this author could find only one where the pickup and magnet could be reversed vertically (Nunan, US4,379,421, 12 Apr 1983; Fig. 7.1).

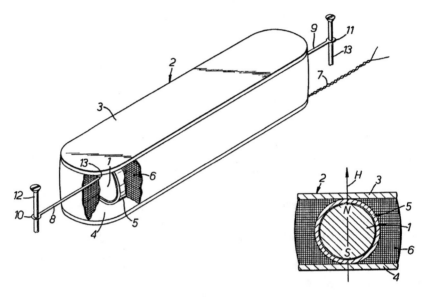

Fig. 7.1 From Figs. 1 and 2, Nunan (1983). The magnet (1) is secured in a tube (5), and the wires (8 and 9) pass through the coil (6), with loops (10 and 11) on the ends to engage the pickup mounting and adjusting screws (12 and 13). To turn the magnet over, the pickup is dismounted from the screws, flipped over, and remounted

Nunan claimed that this design could be used either as a single-coil pickup, or as one-half of a humbucking pickup, which is true enough, but took it no farther. In his Patent Background, he mistakenly stated, "Even when two single-coil pickups are switched for used together, connected to the same preamplifier input, they cannot operate as a hum-bucking arrangement because they both have the same coil/magnet orientation. This limits the range of tonal variety obtainable." Humbucking does not depend upon the orientation of the magnets. The orientation of the magnets only determines whether the signal of a humbucking pair is in-phase (opposite poles up) or out-of-phase (same poles up).

Nunan's tables show only five possibilities for three of these coils, apparently corresponding to a standard 5-way switch, each of the three coils separately, the neck and middle coil together and the middle and bridge coil together. He lists the 2-coil arrangements as "out-of-phase" and humbucking, which contradicts his previous statement. In reality, there are more configurations, each coil separately for 3, plus 6 series and parallel combinations of humbucking pairs, plus 3 humbucking triples.

If, as Nunan suggests, the coil circuits are connected to a preamplifier, the series and parallel combinations will have virtually indistinguishable tones, bringing the total down from 12 to 9, 6 of which are humbucking.

Nunan also made the mistake of giving the impression that preferred embodiment comprised of the use of wires with loops to mount the pickup. Anyone who has worked with wires knows that over time and use this is a flimsy arrangement. He would have been better advised to use plates, like the mounting ears of standard single-coil pickups mounted to the outside cover. His design has features similar to a standard single-coil lipstick, which may have a magnet with a rectangular cross section, except that the lipstick pickup typically has a mounting plate with screw holes fixed to the bottom of the pickup. Merely moving the mounting plate halfway up the pickup would have accomplished Nunan's purpose.

If a set of Nunan's pickups are in a humbucking circuit and one of them is flipped, it changes the polarity of the coil with respect to the external hum field, and the circuit is no longer humbucking. For the circuit to remain humbucking, the coil connections to the rest of the circuit must also be reversed. This is untrue only when the flipped pickup happens to be in a circuit where it does not contribute hum to the output, which can occur in certain symmetrical circuits. Nunan made no provision either for automatically switching pickup terminals or for providing a signal indicating the direction of the flip.

Perhaps as a consequence of these shortcomings, Nunan's design does not appear in the catalogs of at least three of the major US guitar parts companies, or any other place this author has looked, including online. Otherwise, it was a good idea. This chapter will present pickups in which the magnets can be removed and reversed, and one type of pickup which can be flipped (horizontally), but also provides a manually set signal indicating the direction of flip.

In working on a fourth prototype guitar, described in Baker (2016b), the author used generic single-coil pickups with ceramic magnets, and added turns to the coils to get nearly matched hum response. The author reversed the ceramic magnets on two pickups with a strong neodymium magnet, and added small neodymium magnets to the pickups to boost the field, because they were mounted out of sight, underneath the sound board.

Figures 7.2, 7.3, and 7.4 show the kind of pickup mounting setup disclosed in the 2016(b) patent, with all N-up coils, 1 S-up coil, and 2 S-up coils, respectively. Contra-phase humbucking pairs are shown on the left, and in-phase humbucking pairs are shown on the right. Note that all the pairs can be contra-phase for all the pickups with the same pole up, but no pickup arrangement with both N-up and S-up pickups can have all in-phase humbucking pairs.

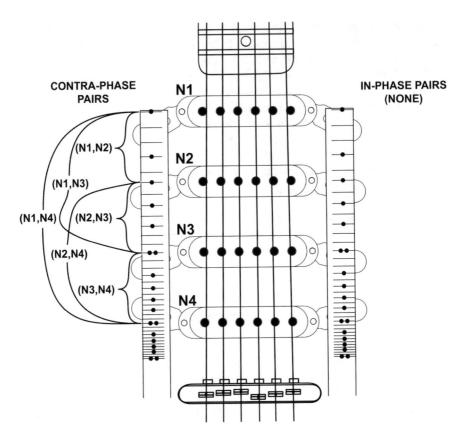

Fig. 7.2 Taken from Fig. 1 in US Non-Provisional Patent Application 15/917,389 (Baker, 2018a), using the pickup mounting system in US Patent 9,401,134 (Baker, 2016b). Virtual fret positions are shown to the left and right of the pickups, with the double-dots at the 36th, 48th, and 60th positions. Showing four N-up pickups mounted at the 24th (2nd octave of fundamental), 29th (5th fret above 2nd octave), 36th (3rd octave), and 48th (4th octave) positions. The bridge shows as a tunematic-style bridge. There is only one pole configuration for all pickups with the same pole up. Note six different humbucking pairs

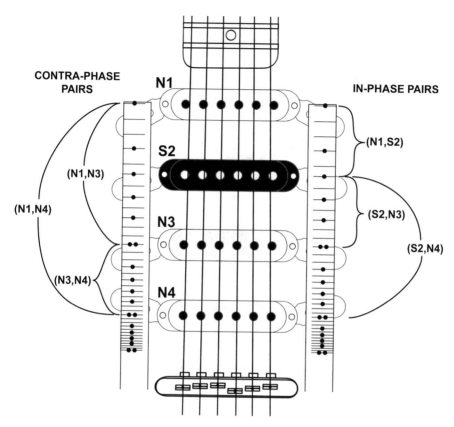

Fig. 7.3 Taken from Fig. 2 in US Non-Provisional Patent Application 15/917,389, showing an S-up pickup (black) replacing an N-up pickup in the second position. Note that all three contra-phase tones have duplicates in Fig. 7.2, and that the number of contra-phase tones equals the number of in-phase tones. There are 4 pole configurations for one pickup with the opposite pole up, one for each pickup position. Note an additional 3 different humbucking pairs, the in-phase (N1, S2), (S2, N3), and (S2, N4) pairs

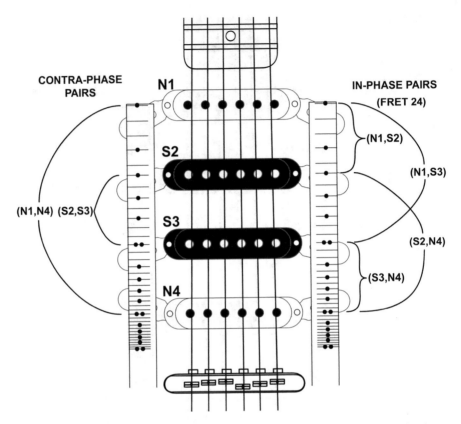

Fig. 7.4 Taken from Fig. 3 in US Non-Provisional Patent Application 15/917,389 showing a second S-up pickup (black) in the third position. Note that the humbucking pair (S2, S3) produces the same tone as an inverted duplicate of (N2, N3) in Fig. 7.2. Note an additional two different humbucking pairs, (N1, S3) and (S3, N4). There are three different pole configurations, (N1, S2, S3, N4), (N1, S2, N3, S4), and (N1, N2, S3, S4). Either (N1, S2, N3, S4) or (N1, N2, S3, S4) will add the remaining different humbucking pair, (N1, S4), for a total of six contra-phase tones and six in-phase tones over all the pole configurations

Note that these figures do not address humbucking triples and quads. The additional examples here, for humbucking triples, quads, and quintets, were developed after filing the patent application, specifically for this book.

7.2   Development of Humbucking Expressions

The following development appears only in this book, not in any patent application. It assumes that the pickup circuit feeds into a high-impedance input amplifier, so that loaded tones need not be considered. We will see that the number of no-load tone circuits, developed from circuit output equations, is generally many fewer than the original number of circuits. In other words, a number of circuits have the same contribution of each pickup to the output, with only a difference in equivalent circuit output impedance. This impedance is expressed in terms of Z_L/Z, where $Z =$ impedance of a single matched sensor, and $Z_L =$ the load impedance which produces an output ½ that of the open-circuit output.

7.2.1 Rules for Humbucking Expressions

In Chap. 4, we applied circuit equations to various circuit topologies and came up with generic expressions for open-circuit output voltage, using the place-holders A, B, C, etc., for pickups, hum signals, and string vibration signals. To simplify and emphasize differences between no-load signals, the expressions for Voc(A, B, C, …) are converted to a fraction with whole numbers in the numerator and a whole number in the denominator, then further simplified by dropping the denominator. In general, we will have an **output expression** like Eq. (7.1), for J number of pickups, where $ci > 0$, $i = 1, \ldots, j$, are the coefficients of the A, B, C, … terms, indicating that the orientation of each pickup in the circuit makes a contribution to the circuit output that is positive, or in-phase with all the other sensors.

$$\text{Voc} \rightarrow c_1 * A + c_2 * B + \cdots + c_j * J \qquad (7.1)$$

To get a **humbucking expression** from an output expression, i.e., Voc $\rightarrow$ 2A + (B + C) transformed to Voc $\rightarrow$ 2A − (B + C), such that A = B = C = Vhum, and Voc $\rightarrow$ 2Vhum − Vhum − Vhum = 0, Eq. (7.2) must be satisfied.

$$\text{Voc} \rightarrow h1 * \text{Vhum} + h2 * \text{Vhum} + \cdots + hj * \text{Vhum} = 0$$

$$h_i = \pm c_i \text{ or } c_i = |h_i|, \quad \sum_{i=1}^{j} h_i = 0, \quad \sum c_i = \text{even} \qquad (7.2)$$

This can happen if and only if the sum of c_i is even, and the signs of h_i can be arranged so that the sum of the positive h_i equals minus the sum of the negative h_i, and equals half the sum of all the c_i. In other words, if the sum of $c_i = 12$, then the sum of the negative h_i must be −6 and the sum of the positive h_i must be 6. That is not possible if the sum of c_i is odd. It would be like trying to divide a pickup in half. We have enough output tonal options this way, that even if it could be done, it is not necessary.

First, imagine that the pickups in Fig. 2.7a, a(2) and b(2), are labeled A and B, each with a matched impedance, Z, grounded on the left with output voltage, Vo, on the right. Then solve the output equations for Vo with a load Z_L across the circuit. Equations (7.3a)–(7.3c) show the results, where Eq. (7.3c) shows the **output expression**. Note that there is only one output expression for both circuits, which differ only in the equivalent circuit output impedance.

$$a(2): \quad \frac{(A + B)Z_L}{2Z + Z_L}, \quad \frac{Z_L}{Z} = 2 \qquad (7.3a)$$

$$b(2): \quad \frac{(A+B)Z_L}{Z+2Z_L}, \quad \frac{Z_L}{Z} = \frac{1}{2} \tag{7.3b}$$

$$Voc \rightarrow A + B \tag{7.3c}$$

Table 7.1 Output expressions for $J = 3$ circuits in Fig. 2.7c, where the A, B, and C columns indicate the c_i coefficients for Eq. (7.1), Z_L/Z indicates the ratio of the equivalent output impedance of the circuit to the matched sensor impedance, "code" is a formula for ordering the relations by the coefficients, "sum coef" is the sum of the coefficients, and HUM/HB shows whether the sum of the coefficients is even (HB) or odd (hum)

#(Type)	A	B	C	Z_L/Z	Code	Sum coef	HUM/HB
a(3)	1	1	1	3	111	3	hum
b(2 + 1)	1	1	2	2/3	211	4	HB
c(2 + 1)	1	1	2	1 1/2	211	4	HB
d(3)	1	1	1	1/3	111	3	hum

Now do the same thing for the $J = 3$ circuits in Fig. 2.7c, as shown above in Table 7.1. Note that there is only one humbucking output expression, as shown in Eq. (7.4).

$$Voc \rightarrow A + B + 2C \tag{7.4}$$

Next, do the same thing for the $J = 4$ circuits in Fig. 2.8. Table 7.2 below shows results similar to Table 7.1, but sorted first by HUM/HB, then by "sum coef," then by "code," with the 3 humbucking output expressions shown in Eq. (7.5). Note especially that the third expression comes from the circuits b(3 + 1) and g(3 + 1), which the methods from Chaps. 2 and 3 in this book would not have identified as humbucking, and Chap. 4 did not cover. And that out of 10 series-parallel circuits, there are only 3 no-load humbucking output expressions for 4 matched pickups.

Table 7.2 Output expressions for $J = 4$ circuits in Fig. 2.8, where the A, B, C, and D columns indicate the c_i coefficients for Eq. (7.1), Z_L/Z indicates the ratio of the equivalent output impedance of the circuit to the matched sensor impedance, "code" is a formula for ordering the relations by the coefficients, "sum coef" is the sum of the coefficients, and HUM/HB shows whether the sum of the coefficients is even (HB) or odd (hum)

#(Type)	A	B	C	D	Z_L/Z	Code	Sum coef	HUM/HB
a(4)	1	1	1	1	4	1111	4	HB
h(4)	1	1	1	1	1/4	1111	4	HB
i(2 + 2)	1	1	1	1	1	1111	4	HB
j(2 + 2)	1	1	1	1	1	1111	4	HB
d(2 + 2)	1	1	2	2	2/5	2211	6	HB
e(2 + 2)	1	1	2	2	2 1/2	2211	6	HB
b(3 + 1)	1	1	1	3	3/4	3111	6	HB
g(3 + 1)	1	1	1	3	1 1/3	3111	6	HB
c(2 + 1 + 1)	1	1	2	3	1 2/3	3211	7	hum
f(2 + 1 + 1)	1	1	2	3	3/5	3211	7	hum

(a) Voc → A + B + C + D

(b) Voc → A + B + 2C + 2D   (7.5)

(c) Voc → A + B + C + 3D

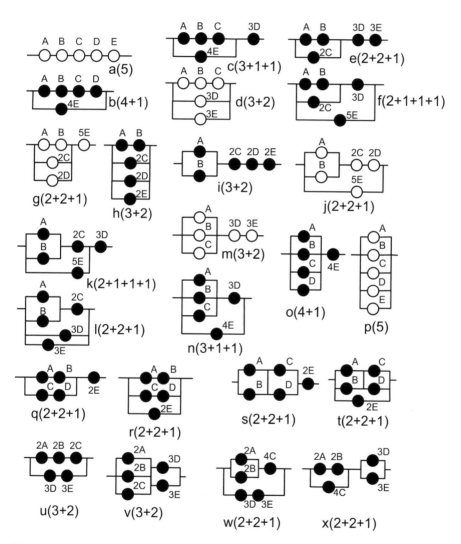

Fig. 7.5 Adapted from Fig. 2.9, and the solutions to the circuit equations for the open circuit voltage, Voc, as described in Sect. 4.2 and Listings 4.1–4.6. The circuits with white circles indicate the 6 circuits that cannot be humbucking, while the black circles indicate the 18 which can. The circles are labeled A, B, C, D, and E as placeholders for pickups and their signals, each preceded by the integer numerator coefficient from the Voc output equation (if greater than 1), dropping the integer denominator. The ground (assumed on the left of the circuit), the output, Voc (assumed on the right of the circuit), and the load resistor between Voc and ground are not shown

Figure 7.5 shows Fig. 2.9 with the pickup circuit symbols labeled according to the position of each pickup, A to E, with the coefficients of the output relations. The circuits with all black symbols are humbucking, and the rest are not. Table 7.3 shows results for the 24 series-parallel circuits for $J = 5$ matched pickups, with the HB and hum results further separated. In this table, the positions for pickups A to E have been resorted so that the coefficients increase from left to right.

Table 7.3 Output expressions for $J = 5$ circuits in Fig. 2.9, where the A, B, C, D, and E columns indicate the c_i coefficients for Eq. (7.1), Z_L/Z indicates the ratio of the equivalent output impedance of the circuit to the matched sensor impedance, "code" is a formula for ordering the relations by the coefficients, "sum coef" is the sum of the coefficients, and HUM/HB shows whether the sum of the coefficients is even (HB) or odd (hum)

#(Type)	A	B	C	D	E	Z_L/Z	Code	Sum coef	HUM/HB
q(2 + 2 + 1)	1	1	1	1	2	2	21111	6	HB
r(2 + 1 + 1)	1	1	1	1	2	1/2	21111	6	HB
s(2 + 2 + 1)	1	1	1	1	2	2	21111	6	HB
t(2 + 2 + 1)	1	1	1	1	2	1/2	21111	6	HB
h(3 + 2)	1	1	2	2	2	2/7	22211	8	HB
i(3 + 2)	1	1	2	2	2	3 1/2	22211	8	HB
b(4 + 1)	1	1	1	1	4	4/5	41111	8	HB
o(4 + 1)	1	1	1	1	4	1 1/4	41111	8	HB
e(2 + 2 + 1)	1	1	2	3	3	2 2/3	33211	10	HB
l(2 + 2 + 1)	1	1	2	3	3	3/8	33211	10	HB
c(3 + 1 + 1)	1	1	1	3	4	1 3/4	43111	10	HB
n(3 + 1 + 1)	1	1	1	3	4	4/7	43111	10	HB
u(3 + 2)	2	2	2	3	3	1 1/5	33222	12	HB
v(3 + 2)	2	2	2	3	3	5/6	33222	12	HB
f(2 + 2 + 1)	1	1	2	3	5	5/8	53211	12	HB
k(2 + 1 + 1 + 1)	1	1	2	3	5	1 3/5	53211	12	HB
w(2 + 2 + 1)	2	2	3	3	4	6/7	43322	14	HB
x(2 + 2 + 1)	2	2	3	3	4	2	43322	14	HB
a(5)	1	1	1	1	1	5	11111	5	hum
p(5)	1	1	1	1	1	2	11111	5	hum
d(3 + 2)	1	1	1	3	3	3/7	33111	9	hum
m(3 + 2)	1	1	1	3	3	2 1/3	33111	9	hum
g(2 + 2 + 1)	1	1	2	2	5	1 2/5	52211	11	hum
j(2 + 2 + 1)	1	1	2	2	5	5/7	52211	11	hum

Equation (7.6) shows the 8 humbucking output expressions, none of which the previous Chaps. 2 and 3 methods would have predicted as humbucking.

$$\text{(a)} \quad Voc \rightarrow A + B + C + D + 2E$$

$$\text{(b)} \quad Voc \rightarrow A + B + 2C + 2D + 2E$$

$$\text{(c)} \quad Voc \rightarrow A + B + C + D + 4E$$

$$\text{(d)} \quad Voc \rightarrow A + B + 2C + 3D + 3E$$

$$\text{(e)} \quad Voc \rightarrow A + B + C + 3D + 4E \qquad (7.6)$$

$$\text{(f)} \quad Voc \rightarrow 2A + 2B + 2C + 3D + 3E$$

$$\text{(g)} \quad Voc \rightarrow A + B + 2C + 3D + 5E$$

$$\text{(h)} \quad Voc \rightarrow 2A + 2B + 3C + 3D + 4E$$

Table 7.4 Humbucking output expressions for $J = 6$ circuits, where the #(type) indicates the category of topology (from copies of research notes to be found at Baker, Feb 2018), A, B, C, D, E, and F columns indicate the c_i coefficients for Eq. (7.1), Z_L/Z indicates the ratio of the equivalent output impedance of the circuit to the matched sensor impedance, "code" is a formula for ordering the relations by the coefficients, and "sum coef" is the sum of the coefficients

#(Type)	A	B	C	D	E	F	Z_L/Z	Code	Sum coef
p(6)	1	1	1	1	1	1	1/6	111111	6
jj(2 + 2 + 2)	1	1	1	1	1	1	2/3	111111	6
jjj(2 + 2 + 2)	1	1	1	1	1	1	2/3	111111	6
lll(2 + 2 + 2)	1	1	1	1	1	1	2/3	111111	6
qqq(3 + 3)	1	1	1	1	1	1	2/3	111111	6
kkk(2 + 2 + 2)	1	1	1	1	1	1	1 1/2	111111	6
mmm(2 + 2 + 2)	1	1	1	1	1	1	1 1/2	111111	6
nnn(3 + 3)	1	1	1	1	1	1	1 1/2	111111	6
a(6)	1	1	1	1	1	1	6	111111	6
fff(2 + 2 + 2)	1	1	1	1	2	2	1/3	221111	8
nn(2 + 2 + 2)	1	1	1	1	2	2	1/3	221111	8
ttt(2 + 2 + 1 + 1)	1	1	1	1	2	2	1/3	221111	8
aaa(4 + 2)	1	1	1	1	2	2	3/4	221111	8
rrr(2 + 2 + 1 + 1)	1	1	1	1	2	2	3/4	221111	8
sss(2 + 2 + 1 + 1)	1	1	1	1	2	2	1 1/3	221111	8
gg(2 + 2 + 2)	1	1	1	1	2	2	3	221111	8
iii(2 + 2 + 2)	1	1	1	1	2	2	3	221111	8
kk(2 + 2 + 2)	1	1	1	1	2	2	3	221111	8
ff((4 + 2)	1	1	2	2	2	2	2/9	222211	10
hhh(4 + 2)	1	1	2	2	2	2	2/9	222211	10
zz(4 + 2)	1	1	2	2	2	2	2/9	222211	10
q(4 + 2)	1	1	2	2	2	2	4 1/2	222211	10
xx(3 + 2 + 1)	1	1	1	2	2	3	4/5	322111	10
ww(3 + 2 + 1)	1	1	1	2	2	3	1 1/4	322111	10
hh(2 + 2 + 1 + 1)	1	1	1	1	2	4	2/3	421111	10

(continued)

Table 7.4 (continued)

#(Type)	A	B	C	D	E	F	Z_L/Z	Code	Sum coef
ll(2 + 2 + 1 + 1)	1	1	1	1	2	4	2/3	421111	10
ii(2 + 2 + 1 + 1)	1	1	1	1	2	4	1 1/2	421111	10
mm(2 + 2 + 1 + 1)	1	1	1	1	2	4	1 1/2	421111	10
b(5 + 1)	1	1	1	1	1	5	5/6	511111	10
o(5 + 1)	1	1	1	1	1	5	1 1/5	511111	10
h(3 + 3)	1	1	1	3	3	3	2/7	333111	12
i(3 + 3)	1	1	1	3	3	3	3 1/3	333111	12
d(4 + 2)	1	1	1	1	4	4	4/9	441111	12
m(4 + 2)	1	1	1	1	4	4	2 1/4	441111	12
l(3 + 2 + 1)	1	1	1	3	4	4	1/3	443111	14
yy(3 + 2 + 1)	1	1	1	3	4	4	1/3	443111	14
e(3 + 2 + 1)	1	1	1	3	4	4	2 3/4	443111	14
t(2 + 2 + 2)	1	1	2	2	5	5	3/7	552211	16
cc(2 + 2 + 2)	1	1	2	2	5	5	2 2/5	552211	16
j(3 + 2 + 1)	1	1	1	3	3	7	2/3	733111	16
g(3 + 2 + 1)	1	1	1	3	3	7	1 3/7	733111	16
pp(3 + 2 + 1)	2	2	2	3	3	6	1/2	633222	18
qq(3 + 2 + 1)	2	2	2	3	3	6	1 5/6	633222	18
dd(2 + 2 + 1 + 1)	1	1	2	2	5	7	3/5	752211	18
s(2 + 2 + 1 + 1)	1	1	2	2	5	7	1 5/7	752211	18
z(2 + 2 + 1 + 1)	1	1	2	3	3	8	5/7	833211	18
w(2 + 2 + 1 + 1)	1	1	2	3	3	8	1 3/8	833211	18
tt(2 + 2 + 1 + 1)	2	2	3	3	4	6	1/2	643322	20
uu(2 + 2 + 1 + 1)	2	2	3	3	4	6	2 1/6	643322	20
v(2 + 1 + 1 + 1 + 1)	1	1	2	3	5	8	5/8	853211	20
aa(2 + 1 + 1 + 1 + 1)	1	1	2	3	5	8	1 5/8	853211	20
ggg(2 + 2 + 2)	2	2	4	4	5	5	8/9	554422	22
bbb(2 + 2 + 2)	2	2	4	4	5	5	1 1/9	554422	22
ddd(2 + 2 + 1 + 1)	2	2	4	5	5	6	1	655422	24
ccc(2 + 2 + 1 + 1)	2	2	4	5	5	6	1 1/9	655422	24

The expressions with hum have been left out as an exercise for the reader

Table 7.4 above shows similar results for the humbucking output expressions for $J = 6$ matched pickups from the circuits in Fig. 7a–e in the research notes illustrated in https://www.researchgate.net/publication/323390784_On_the_Topologies_of_Guitar_Pickup_Circuits (Baker, Feb 2018). From the 72 circuits developed in those research notes, there are 18 no-load humbucking output expressions and 10 no-load output expressions with hum (not shown). The hum expressions and the output relations in the form of Eq. (7.1) are left as an exercise for the reader.

In general, it helps to arrange the letters of the pickups in the circuit so that the absolute magnitude of the coefficients of the output expression increases either from right to left or left to right. For example, the output expressions Voc → A + B + 2C,

Voc → A + B + 2C + 2D, and Voc → A + B + C + 3D + 4E could also be written as Voc → 2A + B + C, Voc → 2A + 2B + C + D, and Voc → 4A + 3B + C + D + E. For $J = 2$ to 5, one can simply take the largest coefficient, add whatever smaller coefficients are necessary to reach (sum of coefficients)/2, and then set the remaining coefficients to the opposite sign. For example, Voc → A + B − 2C, Voc → A − B + 2C − 2D, and Voc → A + B − C + 3D − 4E. For $J = 2$ to 5, this will produce only one answer.

One might think Voc → A + B − 2C + 2D − 2E = (A + B − 2C) ± 2(D − E) an exception, since both terms in parentheses sum to zero, and think that this produces even more choices. But notice that 2C can be exchanged with either 2D or 2E depending upon the ± sign. That means that any "extra" tonal choices are duplicates.

But for $J > 5$, there may be other possibilities. Consider the $J = 6$ output expression in Table 7.4, Voc → A + B + 2C + 3D + 5E + 8F. It has a coefficient sum of 20, so one set of coefficients must sum to +10 and the remaining to −10. The coefficient 8 can sum to 10 by either 8 + 2 or 8 + 1 + 1. Or 8 can be cancelled by 3 + 5, leaving 2 to be cancelled by 1 + 1. If one writes down all the possible cancellations, and picks out the common terms, one gets Voc → (A + B − 2C) ± (3D + 5E − 8F). Note that there can be no exchange of common coefficients between the terms in parentheses in this case. Also, one might expect that for $J \gg 5$, there might be more than two such expressions in parentheses where the coefficients inside the parentheses sum to zero, without any coefficients in common among parentheses.

This kind of math has not yet been taken by this author beyond $J = 6$, which leaves room for contributions by other researchers.

7.2.2 Combinations of Pickups for Each Humbucking Expression

In previous chapters, pickups were switched between positions in the circuit, with the understanding that switching pickups within a basic topology makes no difference to the tone. In this approach, one output expression serves several circuit topologies, and the pickups are switched between coefficients in the expression, which corresponds to switching locations in the related circuits. This author has not yet made any study of whether this violates or confirms any previous rules about switching pickups within a basic topology, leaving it as an exercise for the reader.

For $1 < J = K < 6$, the process is fairly simple. If a humbucking expression has two h_i coefficients of the same sign and magnitude, then switching pickups between them, like switching pickups within a basic topology, has no effect on the tone. After the switch, each pickup still makes the same contribution to the output, so it's a duplicate and doesn't count.

To determine how many different no-load tone circuits a particular humbucking expression can give us, by moving pickups from location to location in the circuit, we count the number of different h_i coefficients, rather like basic topologies in Chap. 4. For example, the humbucking expression for J or $K = 3$, Voc $\rightarrow 2A - B - C$, has 1 coefficient of $(+2)$ and 2 of (-1). That means (3 things taken 1 at a time) times (2 things taken 2 at a time) $= 3 * 1 = 3$. This is confirmed by constructing the humbucking expressions Voc $\rightarrow 2A - B - C$, $2B - A - C$, and $2C - A - B$. For $J = 2$, we have (3 things taken 2 at a time) times (the number of different outputs for $K = 2) = 3 * 1 = 3$. So there are 6 total different tonal combinations of humbucking expressions for $K = 3$, $J = 2$ and 3.

For $J = 4$, the output expressions Voc $\rightarrow A + B + C + D$, Voc $\rightarrow A + B + 2C + 2D$, and Voc $\rightarrow A + B + C + 3D$ produce the humbucking expressions Voc $\rightarrow A + B - C - D$, Voc $\rightarrow A - B + 2C - 2D$, and Voc $\rightarrow A + B + C - 3D$. For Voc $\rightarrow A + B + C - 3D$, we have 1 of (-3) and 3 of $(+1)$, which means (4 things taken 1 at a time) times (3 things taken 3 at a time) $= 4 * 1 = 4$. This could also be written as Voc $\rightarrow A + B + C - 3D$, $A + B + D - 3C$, $A + C + D - 3B$, and $B + C + D - 3A$.

For Voc $\rightarrow A - B + 2C - 2D$, we might think we have 1 of $(+1)$, 1 of (-1), 1 of (2), 1 of (-2), from which we calculate (4 things taken 1 at a time) times (3 things taken 1 at a time) times (2 things taken 1 at a time) times (1 thing taken 1 at a time) $= 4 * 3 * 2 * 1 = 24$. But there's a symmetry of coefficients that cuts that number in half. As in Sect. 4.2.2.3, for every one of the 24 suspected choices, we have a pair of relations like:

$$A - B + 2C - 2D = -(B - A + 2D - 2C) \tag{7.7}$$

The **Rule of Inverted Duplicates** strikes again. A mere reversal of sign at the output cannot produce a tone distinguishable by the human ear, without a separate source of phase reference. There are only 12 different tone combinations of pickups for this for Voc $\rightarrow A - B + 2C - 2D$. The same is true for Voc $\rightarrow A + B - C - D$. We might think we have (4 things taken 2 at a time) times (2 things taken 2 at a time) $= 6 * 1 = 6$, but (moving the coefficients instead of the pickups) there are only $6/2 = 3$: Voc $\rightarrow A + B - C - D$, Voc $\rightarrow A - B - C + D$, and Voc $\rightarrow A - B + C - D$. The reader can verify that the other three expressions are inverted duplicates. We saw this in Sect. 4.2.2.2.

So for $J = 4$, the total number of combinations to produce different no-load humbucking outputs is $4 + 12 + 3 = 19$, not counting combinations of the same pickups with smaller circuits. This, instead of the 15 combinations found in Sect. 4.2.2.5, Eq. (4.5), because of the humbucking $(3 + 1)$ topology.

We now take $K = 4$, and $J = 2$, 3, and 4, to get (4 things taken 2 at a time) times (1 pair) $= 6$, plus (4 things taken 3 at a time) times (3 triples) $= 4 * 3 = 12$, plus 19 for $J = 4$, equals $6 + 12 + 19 = 37$, instead of the 620 series-parallel circuits of ordinary coils in Table 2.4, or the 20 circuits for 4 coils from 2 humbuckers in Table 3.1, or the 21 circuits for matched pickups in Table 4.7.

For $J = K = 5$ matched pickups, Table 7.3 and Eq. (7.5) show 8 different output expressions for humbucking circuits. (We ignore the non-humbucking circuits.) Table 7.5 shows how they can be transformed into humbucking expressions.

Table 7.5 Combinations of pickup circuits for $J = 5$ humbucking expressions

(Sum coef)/2	Output expression	Humbucking expression	Alternate humbucking	Combinations	
				First	Alternate
3	A + B + C + D + 2E	A + B + C − D − 2E	–	10 * 2 = 20	–
4	A + B + 2C + 2D + 2E	A + B + 2C − 2D − 2E	(A + B − 2D) ± (2C − 2E)	10 * 3 = **30**	10 * (3) * 2 = **60**
4	A + B + C + D − 4E	A + B + C + D − 4E	–	5	–
5	A + B + 2C + 3D + 3DE	A + B − 2C + 3D − 3E	(A + B − 2C) ± (3D − 3E)	10 * 3 * 2 = 60	10 * (3) * 2 = 60
5	A + B + C + 3D + 4E	A + B − C + 3D − 4E	(B + 3D − 4E) ± (A − C)	10 * 3 * 2 = **60**	10 * (3 * 2) * 2 = **120**
6	2A + 2B + 2C + 3D + 3E	2A + 2B + 2C − 3D − 3E	–	10 * 1 = 10	–
6	A + B + 2C + 3D + 5E	A − B + 2C + 3D − 5E	(2C + 3D − 5E) ± (A − B)	5 * 4 * 3 * 2 = 120	10 * (3 * 2) * 2 = 120
7	2A + 2B + 3C + 3D + 4E	2A + 2B + 3C − 3D − 4E	(2A + 2B − 4E) ± (3C − 3D)	10 * 3 * 2 = 60	10 * (3) * 2 = 60

Of these 8 output and humbucking expressions for $J = K = 5$ pickups, 5 can be written in more than one way, with an alternate expression composed of two terms in parentheses, each of which sums internally to zero hum. The first humbucking expression has combinations of pickups calculated in the 5th column. The alternate expression, when one exists, has combinations of pickups calculated in the last column. Two of the 5 alternate expressions, in the 2nd and 5th rows below the header, have different combination results.

Consider the humbucking expression Voc $\rightarrow$ A + B − 2C + 3D − 3E. By the primary method presented above, it has 2 coefficients of (+1), 1 of (−2), 1 of (+3), and 1 of (−3). This calculates to combinations of (5 things taken 2 at a time) times (3 things taken 1 at a time) times (2 things taken 1 at a time) times (1 thing taken 1 at a time) = $10 * 3 * 2 * 1 = 60$. The numbers for (1 thing taken 1 at a time) =1, or (2 things taken 2 at a time) = 1, are left out of the table.

The alternate expression, Voc $\rightarrow$ (A + B − 2C) $\pm$ (3D − 3E), can be calculated by (5 things taken 3 at a time) times [(3 things taken 2 at a time) times (1 thing taken 1 at a time)] = $10 * (3 * 1) = 30$, times two things for (3D − 3E) and (3E − 3D), coming to 60, again.

Now consider the humbucking expression Voc $\rightarrow$ A + B − C + 3D − 4E, with 2 of (+1), 1 of (−1), 1 of (+3), and 1 of (−4). Its combinations can be calculated as (5 things taken 2 at a time) times (3 things taken 1 at a time) times (2 things taken 1 at a time) times (1 thing taken 1 at a time) = $10 * 3 * 2 * 1 = 60$. But the alternate expression, Voc $\rightarrow$ (B + 3D − 4E) $\pm$ (A − C), when calculated like Voc $\rightarrow$ (A + B − 2C) $\pm$ (3D − 3E), give us (5 things taken 3 at a time) times [(3 things taken 1 at a time) times (2 things taken 1 at a time) times (1 thing taken 1 at a time)] times (2 things for (A − C) and (C − A)] = $10 * (3 * 2 * 1) * 2 = 120$, twice the number of combinations of the first calculation.

What's the difference?

In Voc $\rightarrow$ (B + 3D − 4E) $\pm$ (A − C), the coefficient of (+1) is in both parentheses. So half of those 120 combinations are duplicates, because any given pair of pickups can be switched between the parentheses on those two coefficients without changing the tone of the output. That makes the alternate calculation of combinations false in this kind of case. This will be true of any parentheses that include the same coefficients.

This means that the total number of pickup combinations for no-load humbucking expressions for $J = 5$ is $20 + 30 + 5 + 60 + 60 + 10 + 120 + 60 = 365$. The enterprising student now has the tools to figure out the number of pickup combinations for the 3 no-load non-humbucking expressions. But in that case, the signs of the coefficients have no meaning other than $2^{J-1} = 16$ phase combinations of $J = 5$ pickups. The coefficients can be treated as being part of the same basic topology if they merely have the same magnitude. The enterprising student can also treat the 8 potentially humbucking output expressions in the same manner, subtract the 365 humbucking combinations, add the combination results for the non-humbucking output relations, and get the total number of non-humbucking no-load combinations for 5 pickups.

Now consider the $J = 6$ output relation Voc $\rightarrow$ A + B + 2C + 3D + 5E + 8F, in which the sum of coefficients is 20, and one possible humbucking expression is Voc $\rightarrow$ A + B − 2C + 3D + 5E − 8F. One can calculate the combinations for 2 of (1), 1 of (−2), 1 of (3), 1 of (5), and 1 of (−8) as (6 things taken 2 at a time) times (4 things taken 1 at a time) times (3 things taken 1 at a time) times (2 things taken 1 at a time) times (1 thing taken 1 at a time) = $15 * 4 * 3 * 2 * 1 = 360$. It can also be written as Voc $\rightarrow$ (A + B − 2C) ± (3D + 5E − 8F), where the calculations can be written as (6 things taken 3 at a time) times [(3 things taken 2 at a time) times (1 thing taken 1 at a time)] times (2 for ±) times [(3 things taken 1 at a time) times (2 things taken 1 at a time) times (1 thing taken 1 at a time)] = $20 * [3 * 1] * 2 * [3 * 2 * 1] = 720$, twice the previous number.

This could be real. The parentheses don't share any coefficients that would create obvious duplicates.

Table 7.6 18 different output expressions for $J = 6$ or $K = 6$, where # is an order number according to the "Sum of Coefficient," Voc $\rightarrow$ is the output expression, "From Topologies" indicates the categories of basic topologies which generated the output expressions, and "Chapter 4 HB Combinations" shows for comparison the combinations generated from circuit categories (6), (4 + 2) and (2 + 2 + 2) calculated from Eqs. (4.6) and (4.7) without the number of versions

#	Voc $\rightarrow$	Sum of coefficients	From topologies	Chapter 4 HB combinations
1	A + B + C + D + E + F	6	(6), (3 + 3), (2 + 2 + 2)	10, –, 360
2	A + B + C + D + 2E + 2F	8	(2 + 2 + 2), (2 + 2 + 1 + 1)	360, –
3	A + B + 2C + 2D + 2E + 2F	10	(4 + 2), (2 + 2 + 2), (2 + 1 + 1 + 1 + 1)	90, 360, –
4	A + B + C + 2D + 2E + 3F	10	(3 + 2 + 1)	–
5	A + B + C + D + 2E + 4F	10	(2 + 2 + 1 + 1)	–
6	A + B + C + D + E + 5F	10	(5 + 1)	–
7	A + B + C + 3D + 3E + 3F	12	(3 + 3)	–
8	A + B + C + D + 4E + 4F	12	(4 + 2)	90
9	A + B + C + 3D + 4E + 4F	14	(3 + 2 + 1)	–
10	A + B + 2C + 2D + 5E + 5F	16	(2 + 2 + 2)	360
11	A + B + C + 3D + 3E + 7F	16	(3 + 2 + 1)	–
12	2A + 2B + 2C + 3D + 3E + 6F	18	(3 + 2 + 1)	–
13	A + B + 2C + 3D + 5E + 7F	18	(2 + 2 + 1 + 1)	–
14	A + B + 2C + 3D + 3E + 8F	18	(2 + 2 + 1 + 1)	–
15	2A + 2B + 3D + 3E + 4E + 6F	20	(2 + 2 + 1 + 1)	–
16	A + B + 2C + 3D + 5E + 8F	20	(2 + 1 + 1 + 1 + 1)	–
17	2A + 2B + 4C + 4D + 5E + 5F	22	(2 + 2 + 2)	360
18	2A + 2B + 4C + 5D + 5E + 6F	24	(2 + 2 + 1 + 1)	–

For $J = 6$ or $K = 6$, Table 7.6 shows the 18 different circuit output expressions. Note that only expression numbers 1, 2, 3, 8, 10, and 17 even have any correspondence to humbucking hextet circuits found in Chap. 4. The rest of the 12 indicated humbucking tone circuits in this table would never have been found by that method.

Just for convenience, let us now replace (K things taken J at a time) with ($K@J$). It is not a standard mathematical expression, but the standard expression of K over J in brackets simply does not fit in a line of text. Here, calculating the number of pickup combinations gets interesting. Expression #1 has already been shown to have only 10 pickup combinations in Sect. 4.2.3, below Table 4.4. And for #6, there is only one possible solution, Voc $\rightarrow$ A + B + C + D + E $-$ 5F, which is $(6@5) = (6@1) = 6$.

But it is possible to write #2 several ways: A + B + C + D $-$ 2E $-$ 2F; A + B $-$ C $-$ D + 2E $-$ 2F; (A + B $-$ 2F) $\pm$ (C + D $-$ 2E); and (A $-$ C) $\pm$ (B $-$ D) $\pm$ (2E $-$ 2F). Depending on what method of calculating combinations one considers valid, they can produce answers from $(6@4) * (2@2) = 15$ for the first expression up to 120 and 240. Consider that Sect. 4.2.2.2 demonstrated that there are only 3 ways to get humbucking expressions from Voc $\rightarrow$ A + B + C + D. So the second expression could be taken as $(6@4) * (3$ ways for A + B + C + D$) * (2$ for 2E $-$ 2F and 2F $-$ 2E$) = 15 * 3 * 2 = 90$.

The value of 90 combinations seems the most reasonable, but how to prove it? One can assign six different orthogonal functions of (t) to A, B, C, D, E, and F, then cycle the functions A(t) to F(t) through all the possible combinations indicated by the 4 trial expressions shown above for #2, collating the sums. There are $6! = 720$ possible function sums for the first and second of the alternate humbucking expressions for #2. For (A + B $-$ 2F) $\pm$ (C + D $-$ 2E), there are 1440 because of the $\pm$ sign. For (A $-$ C) $\pm$ (B $-$ D) $\pm$ (2E $-$ 2F), there are 2880 because of the two $\pm$ sign.

For each of the alternative expressions and methods of creating combinations of pickups, including the internal phase reversals produced by the $\pm$ signs, there will be sets of 15 to 2880 different versions of Eq. (7.8) with combinations of A(t) to F(t) (derived from Eqs. (7.1) and (7.2)).

$$\text{Voc}(t) \rightarrow h_1 * \text{A}(t) + h_2 * \text{B}(t) + \cdots + h_6 * \text{F}(t) \qquad (7.8)$$

Then all one has to do is eliminate all the duplicates, which are either equivalent or equivalent and of the opposite sign. This will likely be a brute-force computer operation, but it will help to eliminate false assumptions and methods, and to clarify the rules for calculating the real number of combinations from these 18 humbucking expressions. Or for that matter, any $J > 5$ number of matched pickups. There is no more time left in the production of this book for this author to do it. It must be left as an exercise for the enterprising reader or researcher, or a later edition of this book.

7.2.3 Summary of Pickup Combinations for Humbucking Expressions (Table 7.7)

Table 7.7 Humbucking tones for $J = 2$ to 5, $K = 2$ to 8, where J is the number of pickups in the humbucking circuit, HB Expr is the number of humbucking output expressions for circuits of J matched pickups, hum Expr is the number of non-humbucking output expressions for J pickups, HB Ckts is the number of pickup combinations for all the humbucking output expressions, K is the total number of pickups available to be switched, Sum of Tones is the number of pickup combinations for all $J \leq K$, and ? indicates that the theory has not yet been fully developed for $J = 6$ and up

$J =$	2	3	4	5	6	
HB Expr	1	1	3	8	18	
hum Expr	0	1	1	3	10	
HB Ckts	1	3	19	365	?	Sum of Tones
K						
2	1					1
3	3	3				6
4	6	12	19			37
5	10	30	95	365		500
6	15	60	285	2190	?	2550+
7	21	105	665	7665	?	8456+
8	28	168	1330	20,440	?	21,966+

7.3 Making Guitars with Multiple Tonal Characteristics

Baker filed a US Provisional Patent Application (PPA) 62/522,487 on June 20, 2017, followed by a Non-Provisional Patent Application (15/917,389, 2018a), then posted an online article[1] to demonstrate why one could want to flip magnets in humbucking circuits of matched single-coil pickups. For K number of single-coil pickups, there are 2^{K-1} possible combinations of N-up and S-up pickups, all offering at least some different tonal selections of in-phase and contra-phase signals. This section reprises and expands that article, followed by a section on pickups constructed to fulfill the promise of guitars with multiple tonal characters. In this chapter, we will consider only the numbers of distinct no-load humbucking circuits, because later chapters will emphasize the use of active circuits. Circuits for loaded tones will be left as an exercise for the reader.

---

[1]Baker, DL, https://www.researchgate.net/publication/323686205_Making_Guitars_with_Multiple_Tonal_Characters, https://doi.org/10.13140/RG.2.2.29053.26081, Mar 2018.

7.3.1 Number of Pole Configurations

Recall from the discussion about Table 2.1 that for J number of pickups, there are 2^{J-1} number of phase changes from reversing individual pickup terminals, where no resulting output is the invert of another. The same argument applies to reversing pickup magnets, because that result is also binary. We note that if an N-up pickup produces a positive signal and an S-up pickup produces a negative signal, then the circuit using S-up pickups S1, S2, and S3, in positions 1, 2, and 3, produces the negative of the signal of the same circuit using N-up pickups, N1, N2, and N3. The same is true of the pickups N1, S2, and N3 versus the pickups S1, N2, and S3.

So, by similar argument, for K number of pickups with reversible magnets, there are 2^{K-1} number of possible pole configurations that don't produce duplicate tones. I.E., for $K = 2, 3, 4$, and 5 pickup with reversible magnets, there are 2, 4, 8, and 16 different pole configurations. We will see that the number of different humbucking tone circuits per pole configuration can share tone circuits with other pole configurations, but that each pole configuration will have a different set.

7.3.2 Signal Phases for Humbucking Pairs with Reversible Magnets

Consider the $K = 2$ output expression, Voc $\rightarrow$ A + B, which has the humbucking expression, Voc $\rightarrow$ A − B. We have only $2^{K-1} = 2^{2-1} = 2$ different pole configurations, one of the inverted duplicates (N_A, N_B) and $(-S_A, -S_B)$, and one of the inverted duplicates $(N_A, -S_B)$ and $(-S_A, N_B)$, where N_A, N_B, $-S_A$, and $-S_B$ represent string signals from pickups A and B. Here we use the convention of N-up producing a "+" string signal, and S-up producing a "−" string signal. We could also consider this to be represented by Voc $\rightarrow$ A $\pm$ B, but then must be careful to keep the signs straight.

7.3.3 Signal and Output Combinations for K = 3 Pickups with Reversible Magnets

Consider the $K = 3$ output expression, Voc $\rightarrow$ 2A + B + C, which has the humbucking expression, Voc $\rightarrow$ 2A − B − C. With reversible magnets, this can be represented by Voc $\rightarrow$ 2A $\pm$ B $\pm$ C, but it is better to keep track of the poles. We have $2^{3-1} = 4$ different pole configurations, as shown in Table 7.8, ignoring the inverted duplicates of $(-S_A, -S_B, -S_C)$, $(-S_A, -S_B, N_C)$, $(-S_A, N_B, -S_C)$, and $(N_A, -S_B, -S_C)$. If the magnets could be reversed with a toggle switch, the number of output choices would also be (3&2 for the number of ways to make a humbucking triple from 2 pickups) times $(2^{K-1}) = 3 * 4 = 12$. Of the 12 different humbucking tones, there are no duplicate or inverted duplicate signals in this case, as confirmed in Table 7.8.

Table 7.8 Physical pole placement of 3 pickups and resulting possible humbucking triples, where N_A, N_B, N_C, $-S_A$, $-S_B$, and $-S_C$ represent string signals for N-up and S-up pickups, A, B, and C, then recast using A, B, and C as string signals

Neck pickup-A	N_A	N_A	N_A	$-S_A$
Middle pickup-B	N_B	N_B	$-S_B$	N_B
Bridge pickup-C	N_C	$-S_C$	N_C	N_C
HB Expressions				
$2A - B - C$ $2B - A - C$ $2C - A - B$	$2N_A - (N_B + N_C)$ $2N_B - (N_A + N_C)$ $2N_C - (N_A + N_B)$	$2N_A - (N_B - S_C)$ $2N_B - (N_A - S_C)$ $-2S_C - (N_A + N_B)$	$2N_A - (-S_B + N_C)$ $-2S_B - (N_A + N_C)$ $2N_C - (N_A - S_B)$	$-2S_A - (N_B + N_C)$ $2N_B - (-S_A + N_C)$ $2N_C - (-S_A + N_B)$
Alternate HB Expressions	$2A - B - C$ $2B - A - C$ $2C - A - B$	$2A - B + C$ $2B - A + C$ $-(2C + A + B)$	$2A + B - C$ $-(2B + A + C)$ $2C - A + B$	$-(2A + B + C)$ $2B + A - C$ $2C + A - B$

Consider in the expression Voc $\rightarrow$ 2A $\pm$ B $\pm$ C. We could also state it as Voc $\rightarrow$ 2X $\pm$ Y $\pm$ Z, so as to allow for placement of A, B, or C in any position. We could also calculate this as $(3@1) * (2@2) * 2^{3-1} = 3 * 1 * 4 = 12$.

Table 7.9 Physical pole placement of 3 pickups and resulting possible humbucking pairs, where N_A, N_B, N_C, $-S_A$, $-S_B$, and $-S_C$ represent string signals for N-up and S-up pickups, A, B, and C, then recast using A, B, and C as string signals, where the **bold** entries are the first non-duplicate of each one, leaving 6 unique with 6 duplicates

Neck pickup-A	N_A	N_A	N_A	$-S_A$
Middle pickup-B	N_B	N_B	$-S_B$	N_B
Bridge pickup-C	N_C	$-S_C$	N_C	N_C
HB Expressions				
$A - B$ $A - C$ $B - C$	$\mathbf{N_A - N_B}$ $\mathbf{N_A - N_C}$ $\mathbf{N_B - N_C}$	$N_A - N_B$ $N_A + S_C$ $N_B + S_C$	$\mathbf{N_A + S_B}$ $N_A - N_C$ $-S_B - N_C$	$-S_A - N_B$ $-S_A - N_C$ $N_B - N_C$
Alternate HB Expressions	$A - B$ $A - C$ $B - C$	$A - B$ $A + C$ $B + C$	$\mathbf{A + B}$ $A - C$ $-B - C = -(B + C)$	$-A - B = -(A + B)$ $-A - C = -(A + C)$ $B - C$

Table 7.9 shows how humbucking pairs ($J = 2$) are made from $K = 3$ matched pickups with reversible magnets. Here we can calculate the total number of different tonal signals as $(3@2) * (2^{J-1}) = 3 * 2 = 6$, as confirmed in the table. There are also 6 duplicates. The number of output choices including duplicates is $(3@2) * 2^{K-1} = 3 * 4 = 12$, because regardless of whether or not one uses 2 or 3 pickups in the circuit, reversing the pickup not being used in a humbucking pair is still an output choice. It's like flipping a 2-throw switch that isn't in the circuit.

Aside 7.1 Missing Denominators—(Aside: something partly relevant one writes down so it won't get lost to the increasing senescence of brain cells) By now, at least one reader may have become terribly annoyed that these "expressions" do not include the denominator. It is not, after all, mathematically correct. But it does save on typing, and makes the overall relationship between coefficients much easier to see. Besides which, have you ever have taken a good look at how a guitar string vibrates? Focus on just one point. Almost all the time, a single point on a steel guitar string will vibrate about its resting axis in a rotating ellipse, of which the minor axis varies between zero and the diameter of a circle. And if you look at the signal from that guitar string created by an electromagnetic pickup, you will see the waveform constantly changing as well, as if the components of the harmonics are constantly changing phase relative to each other. Besides which, the higher harmonics tend to die off faster. In his book on guitars, French (2009) notes that the perceived strength of a tone is very subjective, varying not only with the frequency response curve of human hearing, but with the strength and timing of nearby tones in the signal. Any two guitar players are unlikely to hear the same notes exactly the same way. Then there are the phase effects combined signals from 2 or more pickups, adding and subtracting harmonics in different parts of the signal spectrum. So it is possible that all these effects on the perceived strength of the tone can outweigh the effect of the missing denominator. You can always put it back into your own calculations, when you write your own book.

Aside 7.2 Signals Written as Voltages—One can, if one wishes, write the output expression, Voc → 2A + B + C, in terms string signal voltages, V_A, V_B, and V_C, depending upon the magnetic pole facing the strings. Here, we use the convention that if the pickup is N-up, the signal sign is "+", and if the pickup is S-up, the signal sign is "−". So, knowing that a pickup signal can have either sign with reversible magnets, the proper expression for allowable humbucking string signal voltages over all the pole configurations is Voc → $2V_A \pm V_B \pm V_C$. We do not put ± in front of $2V_A$, because, for example, $-2V_A + (V_B - V_C)$ is merely the negative of $2V_A - (V_B - V_C)$, and inverted duplicate. But at this point, it seems to make a better illustration of what is going on to use +Nx and −Sx for the string signals, to emphasize and keep track of the effects of reversible pickup magnets.

Aside 7.3 Just Electromagnetic Pickups?—Piezoelectric vibration sensors are also bipolar, and can respond to outside electric fields. Hum will not be rejected if two equivalent piezo sensors are placed in the same position on opposite sides of a bending thin plate (sound board). But if one sensor is placed on the thin plate where it bends up, and the other is placed on the same plate where it bends down at the same time, the signal differential between them can reject hum. It will work even better if the entire plate has a grounding layer, especially if the grounding layer sits between the sensors and source of external interference. Keep this in mind when you get to Chap. 11.

7.3.4 Signal and Output Combinations for K = 4 Pickups with Reversible Magnets

For $K = 4$ electromagnetic pickups, we have $2^{K-1} = 8$ different pole configurations. In Sects. 4.2.2 and 7.2.1, and Eq. (7.5), we found 3 different output expressions that can be humbucking and 1 that can't (which we will ignore in this book).

Expression 1 : Voc → A + B + C + D

Table 7.10 Humbucking outputs for the 3 versions of Expression 1, Voc → A + B + C + D, where 8 unique pole configurations are shown in the top 4 rows; positions 1, 2, 3, and 4 correspond to A, B, C, and D; "HB Expressions" show the 3 humbucking versions of the expression, and the columns to the right in that row show the string signals for 3 of the 8 pole configurations; the bottom row represents the corresponding phases of string signals as a binary number with 0 for + and 1 for −; and the bold binary numbers show the first instance of each number from 0000 to 0111

Neck pickup	N1	N1	N1	N1	N1	N1	N1	N1
Upper Middle	N2	N2	N2	N2	−S2	−S2	−S2	−S2
Lower Middle	N3	N3	−S3	−S3	N3	N3	−S3	−S3
Bridge pickup	N4	−S4	N4	−S4	N4	−S4	N4	−S4
HB Expressions								
A + B − C − D	N1 + N2 − N3 − N4	N1 + N2 − N3 + S4	N1 + N2 + S3 − N4	⋮	⋮	⋮	⋮	⋮
A − B + C − D	N1 − N2 + N3 − N4	N1 − N2 + N3 + S4	N1 − N2 − S3 − N4	⋮	⋮	⋮	⋮	⋮
A − B − C + D	N1 − N2 − N3 + N4	N1 − N2 − N3 − S4	N1 − N2 + S3 + N4	⋮	⋮	⋮	⋮	⋮
Phases 0 for + 1 for −	**0011 (3)** **0101 (5)** **0110 (6)**	**0010 (2)** **0100 (4)** **0111 (7)**	**0001 (1)** 0111 (7) 0100 (4)	**0000** 0110 0101	0111 0001 0010	0110 0000 0011	0101 0011 0000	0100 0110 0001

Table 7.10 shows the possible outputs for the 3 humbucking versions of Expression 1, using binary numbers to express the string signal phases for each possible humbucking output. There are 3 humbucking expressions and $2^{4-1} = 8$ possible different pole configurations (out of 16 pole configurations with inverted duplicate pole configurations). If we don't count the inverted duplicate pole configurations as output choices, there are then $3 * 8 = 24$ possible outputs, of which only 8 can be unique tones, leaving 16 duplicates among all the pole configurations.

If the tones are expressed as decimal numbers 0 to 7, the 8 sets contain the tones (3,5,6), (2,4,7), (1,4,7), (0,5,6), (1,2,7), (0,3,6), (0,3,5), and (1,4,6). So we see that in this case, while each of the 8 pole configurations has a unique set of 3 of the 8 tones, the sets are overlapping to the extent that each set can share 0, 1, or 2 tones with any other set, but not all 3. Assuming that the 8 independent tones can be heard to be significantly different, the 8 pole configurations offer 8 different tonal characters, potentially extending the range and versatility of the stringed instrument (guitar or piano, etc.).

For all the pole configurations, we could also state the humbucking expression as Voc $\rightarrow$ W $\pm$ X $\pm$ Y $\pm$ Z, and calculate for 4 coefficients of 1: $(4@4) * 2^{4-1} = 1 * 8 = 8$.

$$\text{Expression 2} : \text{Voc} \rightarrow A + B + C + 3D$$

Just to make things a little simpler, we replace the expression above with Voc $\rightarrow$ 3A + B + C + D, so that the allowable humbucking expressions become 3A – B – C – D, 3B – A – C – D, 3C – A – B – D, and 3D – A – B – C, which we might expect from $(4@1) * (3@3) = 4 * 1 = 4$ combinations. For each of these expressions, there are $2^{4-1} = 8$ different signal tones of the form 3W $\pm$ X $\pm$ Y $\pm$ Z. Note the progression of pickup poles in Table 7.10 from N1, N2, N3, N4 to N1, −S2, −S3, −S4, which can correspond to binary numbers 0000 to 0111, with 0 for a + string signal and 1 for − string signal. So for 4 pickups with 8 pole combinations, this expression gives us $(4@1) * (3@3) * (2^{4-1}) = 4 * 1 * 8 = 32$ different tones, with no apparent duplicates.

$$\text{Expression 3} : \text{Voc} \rightarrow A + B + 2C + 2D$$

The humbucking expressions are of the form W $-$ X $+$ 2Y $-$ 2Z, coming from just 2 versions of the topology category (2 + 2). We might think that we can calculate the number of humbucking expressions from 1 of (+1), 1 of (−1), 1 of (+2), and 1 of (−2) as $(4@1) * (3@1) * (2@1) * (1@1) = 4 * 3 * 2 * 1 = 24$. But consider the first few, staring with: $(A - B + 2C - 2D) = -(B - A + 2D - 2C)$ and $(A - B + 2D - 2C) = -(B - A + 2C - 2D)$. We obviously have inverted duplicates using that approach, so the real answer is 12 different humbucking expressions. We have the first humbucking pair as A – B, A – C, A – D, B – C, B – D, and C – D, then the term $\pm(2Y - 2Z)$ for the remaining 2 pickups. This is $6 * 2 = 12$.

For 8 pole configurations and 12 different humbucking expressions, we have $8 * 12 = 96$ output choices. If we use the expression $Voc \rightarrow W \pm X \pm 2Y \pm 2Z$, we can calculate for 2 of (1) and 2 of (2), $(4@2) * (2@2) * 2^{4-1} = 6 * 1 * 8 = 48$. So far, this calculation has given us the exact number of different humbucking output tones. So the other 48 of 96 tones must be duplicates. Note that if one tries to flip all the magnets into $2^4 = 16$ different pole configurations, the inverted duplicates will then number 144 out of 192.

For all the pole configurations, we have 8 humbucking tone circuits for Expression 1, 32 for Expression 2, and 12 for Expression 3, for a total of 52, compared to the 19 found in Sect. 7.2.2. Even if we expect most of the tones (and the loudest ones) to bunch together at the warm end, this offers a significant increase in choices.

$$Expression\ 4 : Voc \rightarrow A + B + 2C + 3D$$

This last expression cannot be made humbucking, but there is no reason why one cannot apply reversible magnets to it. It becomes $Voc \rightarrow W \pm X \pm 2Y \pm 3Z$, where W to Z are placeholders for moving pickups A to D around in the circuit. It has coefficients 2 of (1), 1 of (2), and 1 of (3). This calculates to $(4@2) * (2@1) * (1@1) * 2^{4-1} = 6 * 2 * 1 * 8 = 96$ different possible tone circuits with hum.

Humbucking combinations for $K = 4, J = 2, 3,$ and 4 with reversible magnets

For $J = 2$, humbucking pairs, the count is $(4@2)$ times the results of $Voc \rightarrow W \pm X$, with 2 coefficients of 1. We get $(4@2) * (2@2) * 2^{2-1} = 6 * 1 * 2 = 12$. For $J = 3$ and $Voc \rightarrow W \pm X \pm 2Y$, we get $(4@3) * (3@2) * 2^{3-1} = 4 * 3 * 4 = 48$. So the count of tone circuits for $2 \leq J \leq K$ is $12 + 48 + 52 = 112$, more than the sum of 37 in Table 7.7.

7.3.5   Method for $Voc \rightarrow h_1X_1 \pm h_2X_2 \pm \cdots \pm h_JX_J$

$$Voc \rightarrow h_1X_1 \pm h_2X_2 \pm \cdots \pm h_JX_J \tag{7.9}$$

Whether humbucking or not, we can now see that calculating the number of tone circuits with reversible magnets is not much different than the method presented in Sect. 7.2. First, set up the output expression as Eq. (7.9). Then, order the coefficients of the expression so that $h_n \leq h_{n+1}$. Say that there are m groupings of coefficients, $1 \leq m \leq J$, so that either h_i is not equal to any other coefficient, or $h_i = h_{i+1} = \cdots = h_{i+n-1}$, where $n > 1$. In other words, gather together all the coefficients h_i into separate groups and count the number in each group. Then figure the number of combinations on that basis, as we have done so many times above, and multiply by 2^{J-1}. It's simple with practice.

This method also has another use. It sets the upper bound on number of possible tone circuits for the humbucking expressions. Even if we have not yet figured out all the calculation rules for humbucking circuit expressions for $K > 5$, we can still calculate the maximum possible number of tone circuits by this method.

7.3.6   Signal Combinations for K = 5 Pickups with Reversible Magnets

Table 7.11 Calculations for the number of tonal circuits for humbucking and non-humbucking circuit output expressions for $K = 5$ matched pickups with reversible magnets, where A, B, C, D, and E are the columns of coefficients for the expressions, Calculations show the calculations of the number of tonal circuits, and Results show the results of those calculations

Humbucking coefficients					Calculations	Results
A	B	C	D	E	$2^{5-1} = 16$	
1	1	1	1	2	$(5@4) * (1@1) * 16 = 5 * 16 =$	80
1	1	2	2	2	$(5@2) * (3@3) * 16 = 10 * 16 =$	160
1	1	1	1	4	$(5@4) * (1@1) * 16 = 5 * 16 =$	80
1	1	2	3	3	$(5@2) * (3@1) * (2@2) * 16 = 10 * 3 * 16 =$	480
1	1	1	3	4	$(5@3) * (2@1) * (1@1) * 16 = 10 * 2 * 16 =$	320
2	2	2	3	3	$(5@3) * (2@2) * 16 = 10 * 16 =$	160
1	1	2	3	5	$(5@2) * (3\&1) * (2@1) * (1@1) * 16 = 10 * 3 * 2 * 16 =$	960
2	2	3	3	4	$(5@2) * (3@2) * (1@1) * 16 = 10 * 3 * 16 =$	480
Total =						2720
Hum coefficients					Calculations	Results
A	B	C	D	E		
1	1	1	1	1	$(5@5) * 16 = 1 * 16 =$	16
1	1	1	3	3	$(5@3) * (2@2) * 16 = 10 * 16 =$	160
1	1	2	2	5	$(5@2) * (3@2) * (1@1) * 16 = 10 * 3 * 16 =$	480
Total =						656

As shown, Table 7.11 calculates the number of possible tonal circuits for $K = 5$ matched pickups with reversible magnets. There are a total of 2720 different humbucking circuit/pickup combinations for the eight $K = 5$ output expressions.

7.3.7   *Signal Combinations for* **K** *= 6 Pickups with Reversible Magnets*

Table 7.12 Calculations for the number of tonal circuits for humbucking circuit output expressions for $K = 6$ matched pickups with reversible magnets, where A, B, C, D, and E are the columns of coefficients for the expressions, Calculations show the calculations of the number of tonal circuits, and Results show the results of those calculations

Humbucking coefficient						Calculations (leaving out $(N@N)$)	Results
A	B	C	D	E	F	$2^{6-1} = 32$	
1	1	1	1	1	1	$(6@6) * 32 = 1 * 32 =$	32
1	1	1	1	2	2	$(6@4) * 32 = 15 * 32 =$	480
1	1	2	2	2	2	$(6@2) * 32 = 15 * 32 =$	480
1	1	1	2	2	3	$(6@3) * (3@2) * 32 = 20 * 3 * 32 =$	1920
1	1	1	1	2	4	$(6@4) * (2@1) * 32 = 15 * 2 * 32 =$	960
1	1	1	1	1	5	$(6@5) * 32 = 6 * 32 =$	192
1	1	1	3	3	3	$(6@3) * 32 = 20 * 32$	640
1	1	1	1	4	4	$(6@4) * 32 = 15 * 32 =$	480
1	1	1	3	4	4	$(6@3) * (3@1) * 32 = 20 * 3 * 32 =$	1920
1	1	2	2	5	5	$(6@2) * (4@2) * 32 = 15 * 6 * 32 =$	2880
1	1	1	3	3	7	$(6@3) * (3@2) * 32 = 20 * 3 * 32 =$	1920
2	2	2	3	3	6	$(6@3) * (3@2) * 32 = 20 * 3 * 32 =$	1920
1	1	2	2	5	7	$(6@2) * (4@2) * (2@1) * 32 = 15 * 6 * 2 * 32 =$	5760
1	1	2	3	3	8	$(6@2) * (4@1) * (3@2) * 32 = 15 * 4 * 3 * 32 =$	5760
2	2	3	3	4	6	$(6@2) * (4@2) * (2@1) * 32 = 15 * 6 * 2 * 32 =$	5760
1	1	2	3	5	8	$(6@2) * (4@1) * (3@1) * (2@1) * 32 = 15 * 4! * 32 =$	11,520
2	2	4	4	5	5	$(6@2) * (4@2) * 32 = 15 * 6 * 32 =$	2880
2	2	4	5	5	6	$(6@2) * (4@1) * (3@2) * 32 = 15 * 4 * 3 * 32 =$	5760
Total =							51,264

As shown, Table 7.12 calculates the number of possible tonal circuits for $K = 6$ matched pickups with reversible magnets. There are a total of 51,264 different humbucking circuit/pickup combinations for the 18 $K = 6$ output expressions.

7.3.8   Summary of Tonal Circuit Combinations of Matched Pickups with Reversible Magnets (Table 7.13)

Table 7.13 Cumulative sums of tonal circuits for $J \leq K$ numbers of matched pickups with reversible magnets, over all the 2^{J-1} pole configurations, where N_K is the number of possibilities for K matched pickups, and the rows below show the combinations for $J \leq K$

$J =$	2	3	4	5	6	
$N_K =$	2	12	52	2720	51,264	
K						Sums
2	2					2
3	6	12				18
4	12	48	52			112
5	20	120	260	2720		3120
6	30	240	780	16,320	51,264	68,634
7	42	420	1820	57,120	358,848	418,250
8	56	672	3640	152,320	1,435,392	1,592,080

7.3.9   Comments on Future Work Needed

This leaves some room for others to make contributions. In particular, all of the calculations in chapters to this point need to be reduced to a computer program and double-checked. First, the program should generate all the possible series-parallel circuit combinations up to some arbitrary number. Second, it should generate all the open-circuit output voltage equations and check the sums of the integer numerator coefficients for possible humbucking circuits. Third, it should generate all the humbucking output relations and collect duplicates. Fourth, it should generate all the possible pickup combinations for each humbucking relation and eliminate inverted duplicates. Fifth, it should complete the calculations to get the numbers of different tone circuits per pole configuration, the number of output selections, and the total number of different tone circuits over all the pole configurations.

7.4   Embodiments of Pickups with Reversible Magnets

To the date of this writing, Jan 2019, this author has not seen any electromagnetic pickups in production that allow the magnet to be reversed. Whether the magnets are alnico or ceramic, are the poles or under them, they are all fixed in position. At most, if the poles are the magnets, they can be raised or lowered by screw threads, but most of those are connected to magnets mounted in or under the coils. All of the drawings in this section come directly from US Non-Provisional Patent Application 15/971,389 (Baker, 2018a). Only the figure numbers have been changed.

7.4.1 Vertical Magnetic Field Pickup Embodiment

Fig. 7.6 Shows the end
view of a single-coil pickup,
with a vertically oriented
magnetic field (toward the
strings), coil (5), magnetic
pole or poles (3), magnet
(1) and a support structure
(2 and 4), in which the
magnet can slip in and out so
that the orientation of the
pole can be reversed

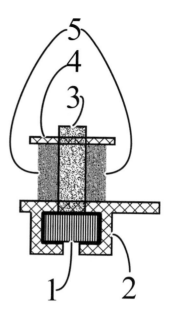

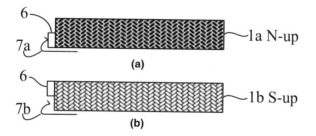

Fig. 7.7 Shows in (**a**) a magnet (1a) with north pole up, with a shorting bar (6) at one end, contacting one or more spring contacts (7a) which send a shorting signal to any guitar pickup switching or electronic control system. In (**b**), it shows the same magnet (1b) with the south pole up, and the shorting bar (6) not contacting one or more spring contacts, so that no shorting signal is sent

Figure 7.6 shows the end-view cross section of an embodiment for a single-coil pickup, modified from the usual type, with upper (4) and lower (2) support structures. The lower structure (2) has 2 L-shaped leg forms that extend downward to enclose the magnet (1) in a U-shaped channel, and center it left-right on the figure, under the pole or poles (3). The pole or poles in this case can be the usual separate rod-shaped poles under each guitar string, or a single bar-shaped pole, either of which can pass through both structures (2) and (4). The magnet is held naturally in position, in-out of the figure, by its magnetic attraction to the pole or poles. The lower support structure (2) has an open channel directly under the magnet to facilitate pushing it out of its resting place with a small probe, such as a toothpick or screwdriver, against the opposing force of its attraction to the poles.

Figure 7.7a shows a view of the magnet with other parts of the pickup removed. Here an electrically shorting bar (6) is attached permanently to one end of the magnet, such that when the north pole is up, or toward the strings, the magnetic attraction force of the magnet (7.7a, b) pushes the shorting bar against a contact or contacts (7a) which then short to each other, to provide a shorting signal to any guitar switching or electronic control system, to signal that the north pole is up. Figure 7.7b shows the same magnet, oriented with the south pole up, and the shorting bar (6) no longer in a position to short the contact or contacts (7b).

The shorting contact or contacts may be positioned or embodied in any number of ways on the pickup support structure, so long as they perform the function as described. They could even use the opposite convention, indicating when the south poles is up, so long as all the pickups on the guitar produce consistent pole orientation signals. Both the lower support structure (2) and the magnet (1) could be metallic, and a single, insulated spring contact (7a–b) could be used to provide a shorting signal to the grounded support structure and magnet. Rare earth magnets are commonly plated with protective and electrically conductive metals. ALNICO magnets are electrically conductive and can be easily constructed with a notch and nub on one end to form the shorting contact (6). Ceramic magnets are typically not electrically conductive, and would need shorting bars attached to them.

Other embodiments or improvements are possible. To hold the magnet in place against sharp bumps to the guitar body, a simple L-shaped spring latch could be added to the opposite end of the magnet channel in the lower support structure (2) to keep the magnet from sliding out that way. The slot in the bottom of the lower support structure could be the full width of the magnet channel, with closed or semi-closed channel ends, to allow the magnet to be easily removed and reversed from the bottom of the pickup, instead of sliding out the end. This would require added latches and/or spring detents to keep the magnet from being bumped out that way. Then the magnet orientation could easily be changed through an access panel in the back of the guitar body, as shown in Baker (2016b, US9,401,134) (i.e., 75 in Fig. 2; 229 in Fig. 18).

In lieu of a shorting bar, the pole orientation sensor could be a light emitter and detector pair, working on reflective and nonreflective portions of the ends of the magnet, or a direct solid-state field detector, such as a Hall effect device. Either of them would need power, and could be used in conjunction with a preamp mounted on the pickup, preferably a fully differential preamp, which would help to reject common-mode noise.

The five paragraphs above come directly from the PPA 62/522,487 (Baker, 2017b) with changes to the numbering of figures and parts. At the time, the pole indicator was considered necessary for a micro-controller or micro-processor switching system, which would not know how to connect matched pickups together into humbucking pairs, quads and hexes, etc., unless it also knew the pole alignments. Since then, further research and consideration has demonstrated that a humbucking circuit does not depend upon the pole orientation, but only upon the canceling connections of hum sources.

Nevertheless, the same body of research has demonstrated, in theory, that if the complex frequency spectrum of each pickup, or each humbucking pair, is known, then the spectrum of any humbucking combination can be calculated by a computational device within or outside the guitar. For this, it may be necessary to know whether the spectrum is to be added (N-up, by the convention here) or subtracted (S-up). Thus, the pole sensor or indicator may still have some relevance.

7.4.2 Horizontal Magnetic Field Pickup Embodiment

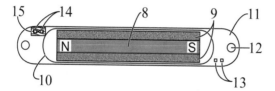

Fig. 7.8 Shows a lipstick-style pickup with a horizontal coil (9) and magnet (8) with left to right poles N-S, a lipstick-style housing (10), a mounting plate (11), with holes (12) for mounting the pickup to the guitar body, header pins at the north end of the magnet (14) that will provide a shorting signal when jumper (15) is attached, and header pins at the south end of the magnet (13) that provide no shorting signal, but a place to keep the jumper when the pickup is reversed and the magnetic poles are S-N, left to right

Note that because the magnetic fields of the vertical (Fig. 7.7) and horizontal (Fig. 7.8) types of pickups are mutually orthogonal, they cannot be mixed in humbucking pair circuits.

Figure 7.8 shows a modified lipstick-style guitar pickup. Instead of the usual vertically oriented magnet and coil, the lipstick body (10) contains a horizontally oriented magnet and coil, with the magnet (8) shown with its north pole to the left, or at the nominal 6-string side. The magnet is a rod or bar, with its length showing in the figure. In this case, the vibrations of the strings passing through the dipole field of the magnet around it, rather than off the end, would produce the signal in the coil (9), wound around the magnet. To provide the shorting signal for detecting pole orientation, two sets of header pins, (13) and (14), are attached to the pickup mounting plate (11).

In the convention shown here, dummy pins (13) sit at the south end of the magnet, merely for holding the shorting jumper (14) so it won't get lost. At least one of the pins (14) is insulated from the mounting plate (11), if it is electrically conductive, so as to provide a shorting signal to any guitar switching or control system when the jumper is placed in that position. If all of the matched pickups on a guitar are of this type, and all of the jumpers are placed on one side, the 1-string side (right in this drawing), or the 6-string side (left in this drawing), then this serves the pole-orientation the signaling function. According to humbucking pairs theory and practice, reversing all of the poles will produce an audio signal indistinguishable from the converse, because there is no other reference signal for the human ear to detect the phase difference.

This embodiment of a lipstick-style pickup has the advantage that to reverse the orientation of the pole, one only needs to detach it from the guitar body, move the jumper from one set of header pins to the other, and reattach the pickup to the guitar body, flipped 180° from before. In this case, the relation to the hum field is also flipped. If the shorting, or other pole orientation, signal is not also used to reverse the string/hum signal terminals, then the pickup will no longer be humbucking in the circuit. Thus, this pole indication feature is necessary.

Figures 7.9, 7.10, 7.11, 7.12, 7.13, 7.14, 7.15, 7.16, 7.17, and 7.18 show experiments with a student 2-D magnetic modeling program.[2] It simulates magnetic fields and materials in two coordinate systems, R–Z cylindrical and X–Y Cartesian. Figures 7.9–7.14 use cylindrical coordinates, which are more realistic.

Using the only resource available to this author, the X–Y coordinate system was necessary but not ideal. The X–Y coordinate system assumes that the material profiles are cross sections of bar and plate forms that extend infinitely into and out of the page. As such, the simulated fields in Figs. 7.15–7.18 do not adequately represent the real fields of the cylindrical objects being modeled.

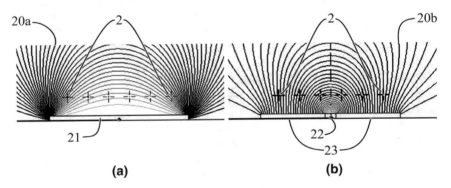

(a)   **(b)**

Fig. 7.9 Shows the magnetic fields (20a (**a**); 20b (**b**)) of two different embodiments (21, and 22, 23) of the horizontal magnetic core, about 0.2″ diameter and 2.5″ long, of the pickup shown in Fig. 7.8. Six guitar strings (2) are shown about 0.45″ above the center of the core. (**a**) Shows the flux lines of a solid ceramic magnet (21), and (**b**) shows the field of a composite magnet with a small rare-earth magnet (22) in the middle of magnetic steel poles (23). The flux lines are more uniformly spaced along the composite magnet

---

[2]Ansoft Corp. (1984–2002). Maxwell, Student Version 3.1.04. Pittsburgh, PA: Ansys. Ansoft Corporation, © 1984–2002 (Acquired by Ansys, 2008). Retrieved from www.ansys.com.

Figure 7.9a, b shows two different embodiments for the magnetic core used in the pickup in Fig. 7.8. Figure 7.9a shows a magnet (21) made of an isotropic material, such as ceramic or Alnico, in rod shape, about 0.2 in. diameter and 2.5 in. long, centered on the bottom horizontal line, indicating the center axis of the magnet. The pickup coil (not shown) would be wrapped around it, with the coil axis the same as the magnet axis. The guitar strings (2), spanning about 2 in., with strings 1 to 6 shown right to left, are shown with dots and crosses about 0.45" above the axis of the magnet. The curvature of the string plane to match fret radius is not shown. A magnetostatic simulation program generated 40 linear flux lines (20a), which have a gray-scale ramp in tone. The lighter lines represent stronger field. The contrast of the image has been enhanced to make the lightest lines show clearly.

Figure 7.9b shows an alternative embodiment, with strings (2) in the same relative position, and a magnetic core (22 and 23) of the same size. In this case, the magnet (22) is smaller, made of rare-earth, centered in the core, and much more powerful than either ceramic or Alnico. The rest of the core (23) are two steel pole pieces of equal size. The same contrast enhancement has been applied to the image to make all of the lighter-gray of the 40 linear flux lines (20b) show up. Note that in this case, the flux lines are distributed almost evenly along the length of the core.

The two parts of Fig. 7.9 show that changing the gradation of the magnetic field along the magnetic core can produce a variety of field distributions, which will affect the tonal character of the pickup output. This raises the possibility of a hotter pickup near the bridge, which can still be matched to the other pickups in response to external hum fields, as well as flux line distributions to produce signals of nearly equal loudness from the vibrations of the strings. Figures 7.10–7.13 show how changing the diameter of the magnet core along its length can affect and adjust the magnetic field.[3]

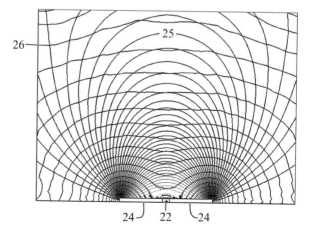

Fig 7.10 Shows the flux lines (25) with a linear distribution, and the magnetic B-field (26) on a logarithmic distribution for a horizontal magnet core with a rare-earth magnet (22) in the center of two Alnico 5 round rods (24)

---

[3]The horizontal-coil pickup is a less efficient design; the string signal output tends to be an order of magnitude less than for the standard vertical-coil design.

Fig. 7.11 Shows the flux lines (25) with a linear distribution, and the magnetic B-field (26) on a logarithmic distribution for a magnet core with a central rare-earth magnet (22) and two tapered rods of Alnico 5, with the end diameters ¼ of the diameter of the central magnet (22). The flux lines are more evenly spaced along the magnet

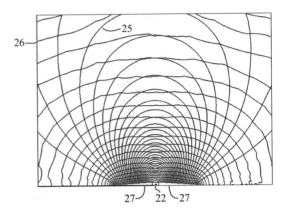

In Figs. 7.10 and 7.11, the magnet core has a central rare-earth magnet (22) with two rods of Alnico 5 (24) attached, magnetized in the same direction. In Fig. 7.10, the Alnico rods are the same diameter as the central magnet, and the linearly distributed flux lines (25) are concentrated at the ends of the magnet core. Consequently, the log-distributed magnetic B-field lines (26) dip sharply in the middle near the core, and in the region of the strings (not shown). But if the Alnico 5 rods (27, Fig. 7.11) are tapered from the diameter of the central magnet (22) to ¼ of the diameter at the outer ends, then the flux lines (25) are more evenly distributed along the magnet core, and the B-field lines (26) are flatter moving away from the core.

Fig. 7.12 Shows the flux lines (28) with a linear distribution, and the magnetic B-field (29) with a logarithmic distribution for a horizontal magnet core (Fig. 9A) made of a round bar of Alnico 5 (30)

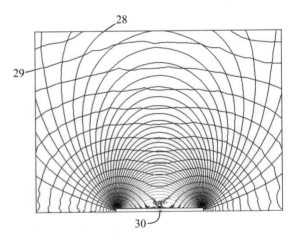

Figures 7.12 and 7.13 show the same kinds of plots of flux lines (28) and magnetic B-field (29) with a solid Alnico 5 rod of constant diameter (30, Fig. 7.12), and a spindle-shaped Alnico 5 rod (31, Fig. 7.13). The spindle-shaped magnetic core had end diameters of ¼ of the central diameter. Figures 7.9–7.13 show that the distribution of magnetic flux and field in the magnetic core of a horizontal magnet pickup (Fig. 7.8) can be set by the gradation of magnetic material

along the axis of the magnet core (Fig. 7.9), and the shape of the magnet core along
the axis (Figs. 7.10–7.13). Therefore it is possible to shape the magnetic field both to
match the extension of fret radius in the string heights and to simulate the effects of
staggered poles in a standard vertical coil pickup.

Fig. 7.13 Shows the same
kind of plot as Fig. 7.12 for a
spindle-shaped Alnico
5 magnet core (31), with the
diameters at the end ¼ of the
center diameter. The flux
lines are more evenly spaced
along the magnet in
Fig. 7.12

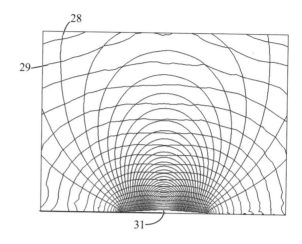

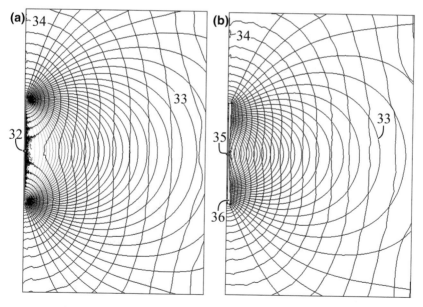

Fig. 7.14 Shows the change in magnetic flux (33) and B-field (34) lines going from a long, thin
rare-earth magnet (32, barely visible in the drawing, (**a**)), with the same volume as the same material
in Fig. 7.9b (22), to a 36% shorter magnet (35), encased in a spindle of cast iron (36), with a
permeability of 60 and an x-shaped diameter profile, smaller in the center than at the ends in (**b**)

Figure 7.14a shows the magnetic flux (33) and B-field (34) lines of a long, thin rare-earth magnet of the same length as all the previous cores, but the same volume of a magnet of the same material (22) in Fig. 7.9b. The magnet itself is barely visible in this drawing, and the field lines are concentrated at the ends, making the magnetic field highly variable across the region where guitar strings might be. Figure 7.14b shows the flux and B-field lines for a magnet (35) of the same material and diameter, but 36% shorter, and encased in a cast iron spindle (36) with a relative permeability of about 60 compared to air or vacuum. The iron spindle has an x-shaped diameter profile, with the center diameter half that of the ends.

In Fig. 7.14, lighter lines indicate higher values of the field, as generated by a magnetic field simulation program. The lines in both 7.14a and 7.14b have been darkened equally to make the lighter lines more visible. It is apparent that the fields are stronger in Fig. 7.14a. Shortening the magnet and encasing it in a cast iron spindle with a diameter that varies linearly along the axis spreads out the flux lines (33) along the composite magnetic core, and flattens the B-field lines (34) in the region where guitar strings might be. This demonstrates that the magnetic field may be manipulated by varying the magnetic properties and dimensions of the magnetic core both radially and axially.

Figures 7.9–7.14 show that the distribution of magnetic flux and field in the magnetic core of a horizontal magnet pickup (Fig. 7.8) can be set by the gradation of magnetic material along the axis of the magnet core (Fig. 7.9), and the shape of the magnet core along the axis (Figs. 7.10–7.14). Therefore it is possible to shape the magnetic field both to match the extension of fret radius in the string heights, and to simulate the effects of staggered poles in a standard vertical coil pickup.

There may be other tradeoffs, in that the magnetic field of a horizontal coil pickups may be more diffuse at the level of the strings than that of a vertical coil pickup, since it is distributed all around the pickup, but using stronger magnetic materials, such as rare-earth magnets, may compensate for this, as well as using an active preamp for each pickup. Coupling effects between nearby pickups, if any, remain to be determined.

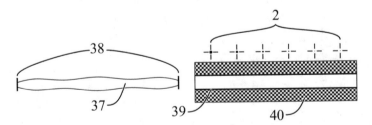

Fig. 7.15 Shows a spindle-shaped prototype magnetic core (37) with nonmagnetic end spacers (38) to maintain its axis in a support tube (39) coincident with the axis of the prototype pickup's wire coil (40). The crosses and dots (2) represent 6 guitar strings, as seen end-on from the bridge. Other structures are not shown

Figure 7.15 depicts both a test setup and a marketable device. As a marketable device, the magnetic core (37, 38) and coil structure (39, 40) can be mounting inside the housing in Fig. 6, which will have end holes and a latching structure (not shown)

to allow the magnetic core to be easily removed, reversed to reverse the poles, to make the changes in Figs. 7.2–7.4. To accomplish that action of the shorting bar, as for the vertical coil magnet in Fig. 7.8, one of the nonmagnetic spacers (38) can be electrically conductive, shorting spring electrical contacts (not shown), while the other spacer is nonconductive.

In a marketable device, the latching structures to keep the magnetic core fixed in the coil may need to be stronger and more positive than for a prototyping device. It can be designed to fit the same mounting holes and space as a standard vertical coil pickup. This is an advantage over many third-party vertical coil lipstick-style pickups, which can be larger and require adapting either the guitar body or pickguard to be mounted.

A test device, Fig. 7.15, can be used in conjunction with a 2-D or 3-D magneto-static field simulation program, rapid prototyping of the core, by methods as simple as turning an Alnico bar on a lathe, and a simulated annealing program. Simulated annealing is a common mathematical computer programming method which operates essentially like finding the deepest pothole in a field of potholes, or conversely the tallest hill in a field of hills. Variables depicting the parameters which can be changed to get to a desire result are defined. In this case, those would include the parameters and profile of a solid of rotation about an axis that defines the shape of the magnetic core (24 and 27 in Figs. 7.10 and 7.11; 30 and 31 in Figs. 7.12 and 7.13; 35 in Fig. 7.14b; 37 in Fig. 7.15). They can also include the variations or gradations of the magnetic properties of the core (21, 22, and 23 in Fig. 7.9; 22, 24, and 27 in Figs. 7.10 and 7.11; 35 and 36 in Fig. 7.14b).

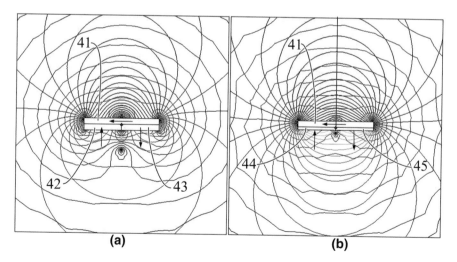

(a)   **(b)**

Fig. 7.16 Shows combinations of a 2-D simulation in the x–y plane of a horizontal alnico field magnet (41) for a horizontal pickup (not shown), with the north pole to the left (as indicated by the long arrow), and rare earth field focusing magnets. (**a**) Shows field focusing magnets with N-up (42) and S-up (43) the same thickness and in direct contract with the alnico main magnet (41), which has N-left. (**b**) Shows an alnico main magnet with field N-left (41), in contact with rare earth focusing magnets, ½ the thickness of the alnico magnet, having N-up on the left (44) and N-down on the right (45)

Figures 7.16 and 7.17 show experimental designs for a more advanced composite magnetic core. A horizontal-field Alnico magnet (41, 46, horizontal arrows), with the north pole to the left, sits above vertical-field focusing magnets, one with its north pole upwards (42, 44, and 47, vertical arrows up) under the horizontal magnet's north pole, and another with its south pole upwards (43, 45, and 48, arrows down). Magnets (41, 42, 43, and 46) are Alnico, while magnets (44, 45, 47, and 48) are rare-earth magnets. Note that the fields above the horizontal magnets are more uniform along the horizontal direction than the fields in Figs. 7.12 and 7.13.

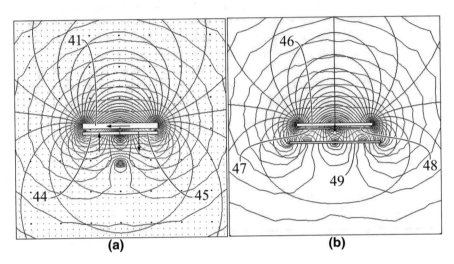

(a)				(b)

Fig. 7.17 Shows combinations of a 2-D simulation in the x–y plane of a horizontal alnico field magnet (41, 46) for a horizontal pickup (not shown), with the north pole to the left (as indicated by the long arrow), and rare earth field focusing magnets (44, 45, 47, 48). (**a**) Shows the same magnets (41, 44, 45) in Fig. 7.16b, but with the focusing magnets (44, 45) separated from the alnico magnet (41) by ½ the thickness of the alnico magnet. (**b**) Shows an alnico main magnet (46) with N-left, above still smaller rare-earth focusing magnets, with N-up on the left (47) and N-down on the right (48), with an iron plate below (49) to limit the extent of the magnetic field below it

Fig. 7.18 Shows a similar arrangement to Fig. 7.17a, but with the focusing magnets (N-up, 51 and N-down, 52) mounted on the iron plate (53) below the alnico main magnet (50)

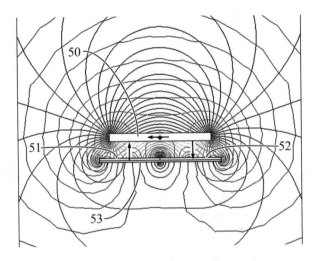

Figure 7.17b shows an iron plate mounted below the pickup coil (not shown), so as to limit the extent of the magnet core field below the pickup. This might reduce any possible effect of a guitar player's ferrous metal belt buckle producing any signal. Figure 7.18 shows the focusing magnets (51 and 52) mounted directly upon the iron plate (53). They seem to have a similar effect on the field of the horizontal magnet (50), as in Fig. 7.16 and 7.17. In the case of Fig. 7.18, putting the focusing magnets on the iron plate makes putting end pins on the magnetic core (50) and reversing it in the coil useless. It would have to be used in the configuration of Fig. 7.8, with the shorting pin set (18) and dummy pin set (17) mounted directly on the pickup body, instead of the ends of the magnetic core.

A 2-D magnetic field simulation program produced each of the drawings, Figs. 7.16, 7.17 and 7.18 in the Cartesian coordinate x–y plane. This means that all the magnets represented as rectangles are modeled as infinitely long ribbons extending into and out of the drawing. This in turn means that all these drawings, as opposed to the others simulated and plotted in the cylindrical coordinate r–z plane, show fields that are only approximations of the 3-D fields that would be produced by 3-D simulation software. Nevertheless, they are useful in trying out designs to see what might work.

Figure 7.17a shows the focusing magnets (44, 45) from Fig. 7.16b shifted out of contact with the main alnico magnet (41), by half the thickness of the alnico magnet. The result is much like 7.16b, but with a slightly flatter B-field in the area of the strings.

Figure 7.17b shows an iron plate mounted below the pickup coil (not shown), so as to limit the extent of the magnet core field below the pickup. This might reduce any possible effect of a guitar player's ferrous metal belt buckle producing any signal. Figure 7.18 shows the focusing magnets (51 and 52) mounted directly upon the iron plate (53). They seem to have a similar effect on the field of the horizontal magnet (50), as in Figs. 7.16 and 7.17.

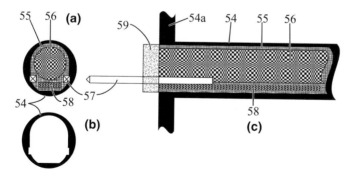

Fig. 7.19 Shows end views ((**a, b**) bobbin cross section alone, for clarity) of a magnet core ((**c**) 55, 56, 57, 58), fitting into a coil bobbin (54, with flanges not shown) with a keyway ((**b**) hole) to maintain proper vertical orientation of the horizontal field magnet (56) and the focusing magnets (58), encased in a nonmagnetic matrix (55). The with end pins (57) are mounted in a piece of printed circuit board (59), and the core matrix (55), to provide both for mating with the keyway, and for shorting pins to indicate the horizontal field direction of the magnet (56)

Figure 7.19a–c shows a composite magnetic pickup core, incorporating the ideas of vertical-field focusing magnets (58), to provide a correction for the horizontal field magnet (56), and end pins (57) to provide a shorting signal when the core in inserted into the coil bobbin (54, with flange 54a) with the north pole in a preferred direction. The end view in Fig. 7.19a shows the horizontal magnet (56) and the focusing magnets (58) are encased and carried in a nonmagnetic matrix (55), as are the end pins (57).

This assembly slides into a keyed hole in the coil bobbin (54), shown a the inner white area of Fig. 7.19b. Figure 7.19c shows the side view of one end of the bobbin (54, 54a), with the magnetic core (55, 56, 57, 58, and 59) inserted. The end pins (57) extend from the matrix (55) through a printed circuit board (59) and out to mate with a female connector (not shown) at one end of the bobbin. The printed circuit at one end of the core shorts the pins, but not at the other end. Thus, when the core is inserted with the north pole in a preferred orientation, the shorted pins produce a shorting signal for any other pickup switching or control circuitry. This is just one possible embodiment. The pins (57) and pc board (59) in Fig. 7.19 can also be used with simpler cores, without focusing magnets, such as those drawn or simulated in Figs. 7.6 and 7.8–7.15.

7.4.3 Comments on Prototype Development

Figures 7.9–7.18 show that magnet material, shape, position, and gradation of field can each be changed to manipulate the resulting field of the pickup magnet. Even asymmetry can be used to induce bumps in the field similar to the staggered poles of a vertical-coil pickup. It remains to put those variations into practice and produce a pickup with an optimum field for a given guitar, pickup position, and style of music. Simulated annealing has been mentioned. Since the time of its introduction, other genetic and evolutionary algorithms have been developed.

Or, it may turn out that a horizontal coil pickup will not work as well one might have hoped. Still, there is a lot of room for other contributions to the field.

References

Baker, D. L. (2016b). Acoustic-electric stringed instrument with improved body, electric pickup placement, pickup switching and electronic circuit, US Patent 9,401,134, 26 July 2016. Retrieved from https://patents.google.com/patent/US9401134B2/

Baker, D. L. (2017a). Humbucking switching arrangements and methods for stringed instrument pickups, US Patent Application 15/616,396, filed 7 June 2017, published as US-2018-0357993-A1, 13 Dec 2018, granted as Patent US10,217,450, 26 Feb 2019. Retrieved from https://www.researchgate.net/publication/335727402_Humbucking_switching_arrangements_and_methods_for_stringed_instrument_pickups_-_NPPA_15616396

Baker, D. L. (2017b). Single-coil pickup with reversible magnet & pole sensor, US Provisional Patent Application 62/522,487, 20 June 2017. Retrieved from https://www.researchgate.net/publication/335727758_Single-Coil_Pickup_with_Reversible_Magnet_Pole_Sensor

Baker, D. L. (2018a). Single-coil pickup with reversible magnet & pole sensor, US Patent Application 15/917,389, 9 Mar 2018. Retrieved from https://www.researchgate.net/publication/331193192_Title_of_Invention_Single-Coil_Pickup_with_Reversible_Magnet_Pole_Sensor

French, R. M. (2009). Engineering the guitar: Theory and practice. New York: Springer. Retrieved from https://www.springer.com/us/book/9780387743684

Nunan, K. N. G. (1983). Electrical pickups, US Patent 4,379,421, 12 Apr 1983. Retrieved from https://patentimages.storage.googleapis.com/33/23/33/a5c5cdae5f54f4/US4379421.pdf; https://patents.google.com/patent/US4379421A/

Chapter 8
Common Connection Point Humbucking Circuits with Odd and Even Numbers of Matched Single-Coil Pickups

8.1 Some Prior Art and Words About Hum

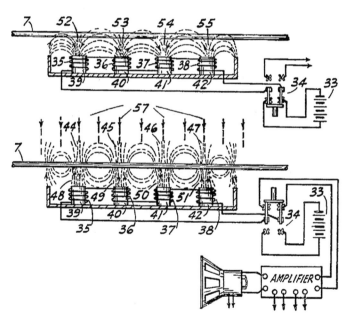

Fig. 8.1 From Figs. 7 and 8, Lesti (US2,026,841, 1936). Shows a switch (34) and battery (33) magnetizing the "steel strings" in Fig. 7, then switching the output to an amplifier in Fig. 8. Note that the coils 35–38 on iron cores 39–42 wind around the cores in alternate directions, providing the cancellation of external interference signals

Lesti (US2,026,841, 1936) may have patented the first humbucking pickup. Figure 8.1 shows his Figs. 7 and 8. They show a switch (34) and battery (33) magnetizing the "steel strings" in Fig. 7, then switching the output to an amplifier in Fig. 8.

© Springer Nature Switzerland AG 2020
D. L. Baker, *Sensor Circuits and Switching for Stringed Instruments*,
https://doi.org/10.1007/978-3-030-23124-8_8

Note that the coils 35–38 on iron cores 39–42 wind around the cores in alternate directions, providing the cancellation of external interference signals. Lesti remarks that permanent magnets would not work because of the distance to the strings, and apparently used soft iron cores which did not magnetize.

Ever since, not a few people have insisted that the coils of humbucking pickups have to be wound in different directions, depending upon the polarity of the magnet involved. This is not so. If one changes the direction of the magnetic field in the pickup, the same thing can be accomplished merely by reversing the connections to the coil. Most coil winding machines wind a right-handed coil. When the coil terminals are reversed, it effectively becomes a left-handed coil. It makes no economic sense to have two separate kinds of winding machine, or even a reversible winding machine.

The only thing that matters is how the pickup coil and poles respond to external hum, not the polarity of the magnetic field, or whether the coil is considered left- or right-handed. If pickups are constructed to have the same hum response, then the pickup terminal connected to the outer windings of the coils, wound the same way on the same machine, will have the same polarity of hum voltage, when the pickups are oriented in the same direction in a uniform hum field. If this is true, then any magnet in any pickup can be reversed, and a humbucking circuit will still be humbucking. The only thing what will change in that case is the relative phase of the string vibration signals between pickups. If the circuit allows the outer windings of all the coils to be grounded, then they have the advantage of acting as an electrostatic shield.

8.2   Simplified Series-Parallel Circuits

The circuits disclosed here accomplish several positive objectives: (1) They introduce an easy method of obtaining humbucking circuits with either odd or even numbers of matched pickups; (2) The open-circuit output voltages all tend to be close to 2 pickups in series, except where the outputs include contra-phase signals; (3) The circuit connections are very simple, with each pickup connected between a common connection point, which can be grounded, and one of two output terminals, making then easy to switch; and (4) If the common connection point is grounded, the matched pickups can be constructed so that the outer windings of every one of them are grounded, to form an electrostatic shield. In return, the negative trade-off results in fewer different tone circuits than were developed for similar numbers of pickups in Chap. 7.

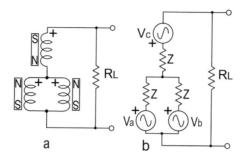

Fig. 8.2 From Fig. 13 in Baker (2017a); a humbucking triple, constructed of matched single-coil pickups, of one S-up pickup in series with a parallel pair of N-up pickups (**a**, with string signal phases shown), showing the hum signals, Vhum = Va = Vb = Vc, and matched impedances, Z, in (**b**). The "+" signs in (**a**) show the polarity of the string voltage, and the polarity of hum voltage in (**b**)

Baker (2017a) identified what was in that patent application a special case, a humbucking triple, as shown in Fig. 8.2. Later, Baker (2018b) developed a more general equation. But this still included the unnecessary string signals, and can be even more simply expressed with hum voltages alone, as in Fig. 8.3 and Eq. (8.1). If the hum voltages cancel at the output, then the circuit is humbucking, and in this circuit, $(m + n)$ can be either odd or even, for $m \geq 1$, $n \geq 1$.

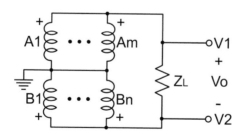

Fig. 8.3 General matched-pickup humbucking circuit, in which all of the coils, $m > 0$, $n > 0$, and $m + n > 1$, are connected to a common connection point (grounded in this representation) by the terminals with the same hum polarity, opposite of that shown as "+." All of the Ai coils are connected to one output terminal, V1, and all of the Bi coils are connected to the other output terminal, V2, across which a load Z_L is placed, for calculating the lumped circuit impedance, Zc, the output voltage being Vo = V1 − V2

(a) $\dfrac{V1 - V2}{Z_L} + \displaystyle\sum_{i=1}^{m} \dfrac{V1 - V_{Ai}}{Z} = 0$

(b) $\dfrac{V2 - V1}{Z_L} + \displaystyle\sum_{i=1}^{n} \dfrac{V2 - V_{Bi}}{Z} = 0$

(c) $\mathrm{Vo} = V1 - V2$

(d) $\mathrm{Voc} = \displaystyle\lim_{Z_L = \infty} [\mathrm{Vo}]$ (8.1)

(Solution) $\mathrm{Voc} = \dfrac{n \displaystyle\sum_{i=1}^{m} V_{Ai} - m \displaystyle\sum_{i=1}^{n} V_{Bi}}{mn}$, $\dfrac{Zc}{Z} = \dfrac{m+n}{mn}$

If $V_{Ai} = V_{Bi} = \mathrm{Vh}$, hum voltage

Then $\mathrm{Vo} = \dfrac{nm\,\mathrm{Vh} - mn\,\mathrm{Vh}}{mn} = 0$

Voc is the open-circuit output voltage, where Z_L goes to infinity. This equation can also be used for signal voltages by assuming a simple convention. Although string vibration signals are not related to hum signals, let the N-up pickups be considered to have the same phase as the hum voltage, and the S-up pickups be assumed to have the opposite phase. Then for N-up pickups, their signal voltages in Eq. (8.1) maintain a positive sign. But for S-up pickups, their signal voltages must have a negative sign. This convention is just a convention and will work just as well reversed, but why complicate things? So if the Ai string signals come from pickups N1 and S1, and the Bi signals come from N2 and S2, for $m = n = 2$, then Voc $= (2(N1 - S1) - 2(N2 - S2))/4 = (N1 - N2 + S2 - S1)/2$.

As the development of humbucking pair theory in Chap. 3 shows, any humbucking circuit in series or parallel with another humbucking circuit is still humbucking. But this chapter will only cover the circuits possible with Fig. 8.3. There is another set (not presented in this chapter) where the Ai pickups are connected in series across Vo in one polarity, and all the Bi pickups are connected in series across Vo in the opposite polarity. For the same Ai and Bi, both circuits produce the same no-load tone. The other set tends to give higher outputs and lumped internal impedances, with $Zc/Z = mn/(m+n)$.

The humbucking circuits in Fig. 8.3 can be said to follow Four Simple Rules: (1) all of the pickups or sensors are connected to a common point at the pickup terminals that present the same phase of external electromagnetic interference, or "hum"; (2) at least two pickups must be in the circuit, connected at least one from the common point to the output low terminal, and at least one connected from the common point to the output high terminal; and (3) either the common point can be grounded, or the low terminal of the output can be grounded, but not both; and (4) the pickups or sensors must be matched, all having the same impedance and the same response to external hum.

Note that with active circuits including amplifiers, one or the other of the common point to lower output terminal should be grounded to keep the output signal from drifting far enough from the ground reference to cause clipping. Usually passive circuits are grounded at the output. The advantage of grounding the common connection point allows the active amplifier connected to the pickup circuit to be differential, with both inputs referenced to ground. This tends to eliminate another form of external signal interference, called common-mode interference, which can come from higher-frequency sources, like fluorescent lights.

8.2.1 Common-Connection Circuits with Two Coils

With any two coils, (N1, N2), (N1, S1), or (S1, S2), indicating the available coils with either N-up or S-up fields, one coil connects to the high output terminal and the other to the low output terminal. Let the first number represent the upper coil and the second the lower coil. This produces the signals (N1 − N2), (N1 + S1), and (−S1 + S2). Reversing those connections only changes the sign of the output signal. This inventor contends that this produces no effective difference in tone. Human ears cannot tell the differences in the phase of a signal producing a tone without some other external reference. Therefore, such changes do not count.

As first noted here in Sect. 4.2.4, the reduction of the number of unique circuits by a factor of 2, due to vertical and horizontal topological symmetry can happen whenever the number of pickups connected to the high and low output terminals is equal to half the total. The number calculated by combinatorial math must be divided by 2. So what might have been (2 things taken 1 at a time) ∗ (1 thing taken 1 at a time) = 2 ∗ 1 = 2, becomes 2/2 = 1. As noted in Chap. 7, the magnet must be reversed to produce the opposite phase of string signal and tone.

8.2.2 Common-Connection Circuits with Three Coils

Suppose that the three coils can be represented by the designations N1, S1, and N2, for 1 S-up and 2 N-up coils. They can be connected through the switching system to the output terminals as either 2 coils or 3 coils. For three coils there can be only 1 coil connected to one output terminal and 2 coils connected to the other, in combinations of (3 things taken 1 at a time) ∗ (2 things taken 2 at a time) = 3 ∗ 1 = 3. Table 8.1 shows various possible circuit/switching combinations. Note that reversing the output terminals produces the duplicates in the right three columns of the table. It does not matter if the circuits are switched this way; it only matters that duplicates are not counted as separate circuits and possible tones. This might be called the Fifth Simple Rule, but it might wait until actual human trials are conducted to confirm it. Call it instead the Rule of Inverted Duplicates.

Note that in Table 8.1, for 2 coils, the results for 2 coils can be explained as (3 things taken 1 at a time) times the number of combinations for 2 coils, or $3 * 1 = 3$. The results for 3 coils can be taken as (3 things taken 1 at a time) * (2 things taken 2 at a time), or $3 * 1 = 3$. The combined results for 3 coils, taken in pairs and triples, is 6 humbucking circuits. By Eq. (8.1), for the first column of 2 coils, $Vo = V_{N1} + V_{S1}$, for the first column of 3 coils, $Vo = V_{N1} + (V_{S1} - V_{N2})/2$, and for the second column of 3 coil duplicates, $Vo = (V_{N1} + V_{N2})/2 + V_{S1}$. The Rule of Inverted Duplicates also applies to any reversal of all the magnetic poles at once.

Table 8.1 Circuit/switching combinations for three coils, N1, S1, and N2, with upper coils connected from the common connection point to the high output terminal, and lower coils connected from the common point to the low output terminal. Note that the pickup designation also stands in for the string signal voltage

				Duplicates		
2	N1	N1	S1	S1	N2	N2
Coils	S1	N2	N2	N1	N1	S1
3	N1	S1	N2	S1N2	N1N2	N1S1
Coils	S1N2	N1N2	N1S1	N1	S1	N2

Table 8.2 Circuit/switching combinations for three N-up coils, N1, N2, an N3, with upper coils connected from the common connection point to the high output terminal, and lower coils connected from the common point to the low output terminal

2 Coils			3 Coils		
N1	N1	N2	N1	N2	N3
N2	N3	N3	N2N3	N1N3	N1N2

It still works for all pickups N-up, N1, N2, and N3, as shown in Table 8.2, shown without the duplicates. By Eq. (8.1), the first column of 2 coil combinations has an output voltage of $Vo = V_{N1} - V_{N2}$. The first column of 3 coil combinations has an output voltage of $Vo = V_{N1} - (V_{N2} + V_{N3})/2$.

8.2.2.1 Common Connection Point Circuits for $K = 3$ Matched Pickups with Reversible Magnets

Now, reboot and consider the case of pickups with reversible magnets, with coils in positions 1, 2, and 3: (N1, N2, N3), (S1, N2, N3), (N1, S2, N3), and (N1, N2, S3). As in Chap. 7.3.3, the 4 pole configurations for 3-coil circuits each have $4 * 3 = 12$ output selections, and 12 different choices of humbucking-triple tone circuits. For 2-coil circuits, there are (3 things taken 1 at a time) times (2 things taken 1 at a time) times (1/2 for symmetry) = 3 tone circuits for each pole configuration, and $4 * 3 = 12$ output selections. But as in Sect. 7.3.2, there are only 6 different tone circuits over all the pole configurations. So, counting both 2-coil and 3-coil circuits using the common-connection humbucking circuit, there are 6 tone circuits for each pole configuration, out of 18 over all the pole configurations, as shown in Table. 7.10.

8.2.3 Common-Connection Circuits with Four Coils

Suppose that we have four matched pickups designated N1, S1, N2, and S2. We can calculate the number of possible outputs for pairs and triples by taking 4 things 2 at a time and 4 things 3 at a time, multiplied by the number of possible pairs (1) and triples (3) without extra pickups. Equation 8.2 shows this calculation.

$$
\text{Pairs from 4 coils:} \quad \binom{4}{2} * 1 = \frac{4 * 3}{2 * 1} * 1 = 6
$$

$$
\text{Triples from 4 coils:} \quad \binom{4}{3} * 3 = \frac{4 * 3 * 2}{3 * 2 * 1} * 3 = 12
$$

(8.2)

For four matched pickups, the circuit in Sect. 7.3.6 and Table 7.6 does not fit the common-connection point model. There are 2 ways remaining to arrange 4 coils in a humbucking quad: (1) a single coil in series with (or over) 3 coils in parallel, and (2) 2 coils in parallel, the pair in series with (or over) another 2 coils in parallel. Putting 3 coils in parallel over 1 coil would merely duplicate the first instance by the Rule of Inverted Duplicates. This will be true for any number of pickups J. If we follow the convention of putting the smaller number of pickups over the larger or equal, the number of pickups connected to the high output terminal will range from 1 to $J/2 - 1$ for J odd, and 1 to $J/2$ for J even. Table 8.3 shows the switched combinations for $J = 4$, given 2 N-up pickups N1 and N2, and 2 S-up pickups, S1 and S2.

Table 8.3 Switching/combinations for 4 coils, N1, S1, N2, and S2, using the conventions that the bottom coils feed into the negative (or low) output, the upper coils feed into the positive (or high) output, the N-up coils have positive string signals, and the S-up coils have negative string signals

1 over 3	N1 N2S1S2	N2 N1S1S2	S1 N1N2S2	S2 N1N2S1	
Vo =	$V_{N1} +$ $(V_{S1} + V_{S2} - V_{N2})/3$	$V_{N2} +$ $(V_{S1} + V_{S2} - V_{N1})/3$	$-V_{S1} +$ $(V_{S2} - V_{N1} - V_{N2})/3$	$-V_{S2} +$ $(V_{S1} - V_{N1} - V_{N2})/3$	
				Duplicates	
2 over 2	N1N2 S1S2	N1S2 S1N2	S1N2 N1S2	S1S2 N1N2	N2S2 N1S1
Vo =	$(V_{N1} + V_{N2})/2 +$ $(V_{S1} + V_{S2})/2$	$(V_{N1} - V_{S2})/2 +$ $(V_{S1} - V_{N2})/2$	$(V_{N2} - V_{S1})/2 +$ $(V_{S2} - V_{N1})/2$	$(-V_{S1} - V_{S2})/2 +$ $(-V_{N1} - V_{N2})/2$	$(V_{N2} - V_{S2})/2 +$ $(V_{S1} - V_{N1})/2$
Duplicates	N1S1 N2S2				
Vo =	$(V_{N1} - V_{S1})/2 +$ $(V_{S2} - V_{N2})/2$				

Of the circuits developed for humbucking quads in Sect. 4.2, only Fig. 4.14.c (4) fits in this more limited set, as shown in the 2-over-2 examples. The number of 2-over-2 combinations can be calculated as one-half times (4 things taken 2 at a time) times (2 things taken 2 at a time), or $6 * 1/2 = 3$. The number of 1-over-3 combinations can be calculated as (4 things taken 1 at a time) times (3 things taken 3 at a time), or $4 * 1 = 4$, for a total of 7 humbucking circuits from 4 pickups. So for 4 pickups, there are $6 + 12 + 7 = 25$ unique common-connection circuits made of 2, 3, and 4 pickups. Note that when all the terms are collected for the 2-over-2 circuits, Vo for the duplicates is the negative of Vo for the first three, due again to the Rule of Inverted Duplicates. This will happen whenever $j = k$ for j-over-k circuits.

8.2.3.1 Common Connection Point Circuits for $K = 4$ Matched Pickups with Reversible Magnets

Consider Eq. (8.2). For pickups with reversible magnets, the number of output selections for humbucking pairs with 8 pole configurations $6 * 8 = 48$. For humbucking triples, it is $12 * 8 = 96$ output selections. For humbucking quads here, it is $4 * 8 + 3 * 8 = 56$, for a total of 200 output selections.

For humbucking pairs, (A, B), over all the pole configurations, the output relation for phase combinations is $Voc \rightarrow V_A \pm V_B$, giving 2 combinations times (4 things taken 2 at a time) $= 2 * 6 = 12$ different no-load tone circuits. For humbucking triples, (A, B, C) the open-circuit output relation for phase combinations is still $Voc \rightarrow 2V_A \pm V_B \pm V_C$. There are (4 things taken 3 at a time) times (3 things taken 1 at a time) times $(2^{3-1}) = 4 * 3 * 4 = 48$ different no-load tone circuits. For humbucking quads in a common-point connection switching system, the output relations for phase combinations are $Voc \rightarrow V_A \pm V_B \pm V_C \pm 3V_D$ and $Voc \rightarrow V_A \pm V_B \pm V_C \pm V_D$. The first produces (4 things taken 3 at a time) times $(2^{4-1}) = 4 * 8 = 24$ different no-load tone circuits. The second produces (4 things taken 4 at a time) times (2^{4-1}) $= 1 * 8 = 8$ different no-load tone circuits, for a total of 32. Adding the pairs and triples, there are $12 + 48 + 32 = 92$ different no-load tone circuits over all the pole configurations for $K = 4$ matched single-coil pickups in a common-point connection switching circuit.

8.2.4 Circuits with Five Coils

For 5 coils, one can take the previous numbers of tonal circuits calculated for 2, 3, and 4 coils and multiply them by 5 things taken 2, 3, and 4 at a time. Unique combinations of 5 coils or pickups in this switching system can be "quint" combinations of 1-over-4 and 2-over-3, without duplicate inversions. Equation 8.3 shows these calculations; note that 5 pickups together have 15 unique combinations.

Pairs from 5 coils: $\begin{pmatrix} 5 \\ 2 \end{pmatrix} * 1 = 10 * 1 = 10$

Triples from 5 coils: $\begin{pmatrix} 5 \\ 3 \end{pmatrix} * 3 = 10 * 3 = 30$

Quads from 5 coils: $\begin{pmatrix} 5 \\ 4 \end{pmatrix} * 7 = 5 * 7 = 35$ (8.3)

$\frac{1}{4}$ Quints from 5 coils: $\begin{pmatrix} 5 \\ 1 \end{pmatrix} * \begin{pmatrix} 4 \\ 4 \end{pmatrix} = 5 * 1 = 5$

$\frac{2}{3}$ Quints from 5 coils: $\begin{pmatrix} 5 \\ 2 \end{pmatrix} * \begin{pmatrix} 3 \\ 3 \end{pmatrix} = 10 * 1 = 10$

Total tonal circuits from 5 coils $= 10 + 30 + 35 + (5 + 10) = 90$

8.2.4.1 Common Connection Point Circuits for $K = 5$ Matched Pickups with Reversible Magnets

Five-pickup circuits using matched single-coil pickups with reversible magnets can have $2^{5-1} = 16$ different pole configurations. That means this kind of circuit can have $16 * 10 = 160$ output selections for humbucking pairs, $16 * 30 = 480$ output selections for triples, $16 * 35 = 560$ output selections for quads, and $16 * 15 = 240$ output selections for humbucking quints. This totals to $160 + 480 + 560 + 240 = 1290$ output selections over all 16 pole configurations.

The number of different tone-circuits for pairs, triples, and quads will be (5 things taken 2 at a time) times (2 phase combinations) $= 10 * 2 = 20$ for pairs; (5 things taken 3 at a time) times (3 things taken 1 at a time) times $(2^{3-1}) = 10 * 3 * 4 = 120$ different tone-circuits for triples; and (5 things taken 4 at a time) times $(24 + 8) = 5 * 32 = 160$ different tone circuits for quads. By Eq. (8.1), the two phase combination relations for quints are Voc $\rightarrow V_A \pm V_B \pm V_C \pm V_D \pm 4V_E$ for 1-over-4 quints, and Voc $\rightarrow 3V_A \pm 3V_B \pm 2V_C \pm 2V_D \pm 2V_E$ for 2-over-3 quints. The first one produces (5 things taken 1 at a time) times $(2^{5-1}) = 5 * 16 = 80$ different tone circuits. The second produces (5 things taken 2 at a time) times $(2^{5-1}) = 10 * 16 = 160$ different tone circuits, for a total of 240 different tone circuits for quints over all the pole configurations. Adding that with the tone circuits for pairs, triples, and quads means $20 + 120 + 160 + 240 = 540$ different humbucking no-load tone circuits over all the 16 pole configurations.

Each of the 16 pole configurations have a set of 90 different tone circuits, out of a pool of 540 different tone circuits, with $1290 - 540 = 750$ duplicate output selections, all humbucking. According to Table 2.4 in Sect. 2.8, the number of humbucking and hum-producing series-parallel circuits for $K = 5$ and $J = 1$–5 is 10,717. This should, of course, be independently checked and verified.

One guitar maker promotes their commercially available electric guitar with 2 humbuckers and a single as having "over 250,000" different pickup circuits or tones. To the best of this author's knowledge, when the last web site was checked, it offered no proof of that claim, nor any analysis as to how many of this huge number of tonal selections may be humbucking, hum-producing or tonal duplicates.

8.2.5 Circuits with Six Coils

A number of guitars on the market have three humbuckers, which can be considered 6 matched pickups for this discussion. Equation 8.4 shows these calculations. Note the reduction of 3-over-3 hextets due to the Rule of Inverted Duplicates.

$$\text{Pairs from 6 coils:} \quad \binom{6}{2} * 1 = 15 * 1 = 15$$

$$\text{Triples from 6 coils:} \quad \binom{6}{3} * 3 = 20 * 3 = 60$$

$$\text{Quads from 6 coils:} \quad \binom{6}{4} * 7 = 15 * 7 = 105$$

$$\text{Quints from 6 coils:} \quad \binom{6}{5} * 15 = 6 * 15 = 90$$

$$\frac{1}{5} \text{ Hexes from 6 coils:} \quad \binom{6}{1} * \binom{5}{5} = 6 * 1 = 6$$

$$\frac{2}{4} \text{ Hexes from 6 coils:} \quad \binom{6}{2} * \binom{4}{4} = 15 * 1 = 15$$

$$\frac{3}{3} \text{ Hexes from 6 coils:} \quad \frac{1}{2}\binom{6}{3} * \binom{3}{3} = \frac{1}{2} 20 * 1 = 10$$

$$\text{Total tonal circuits from 6 coils} = 15 + 60 + 105 + 90 + (6 + 15 + 10) = 301$$

$$(8.4)$$

8.2.5.1 Common Connection Point Circuits for $K = 6$ Matched Pickups with Reversible Magnets

Six matched pickups have $2^{6-1} = 32$ different pole configurations. From Eq. (8.4), the output selections for humbucking circuits of 2, 3, 4, 5, and 6 pickups number $15 * 32 = 480$, $60 * 32 = 1920$, $105 * 32 = 3360$, $90 * 32 = 2880$, and $31 * 32 = 992$ tone circuits, respectively. Adding all the circuits from $J = 2$ to 6 gives $480 + 1920 + 3360 + 2880 + 992 = 9632$ output selections.

Extending previous results, the number of different tone circuits over all the 32 pole configurations for $K = 6$ and $J = 2, 3, 4,$ and 5 are (6 things taken 2 at a time) times $2 = 30$, (6 things taken 3 at a time) times $12 = 240$, (6 things taken 4 at a time) times $56 = 840$ and (6 things taken 5 at a time) times $240 = 1440$, respectively.

The phase combination relation for 1-over-5 coils (F over A to E) is $Voc \rightarrow V_A \pm V_B \pm V_C \pm V_D \pm V_E \pm 5V_F$, giving (6 things taken 5 at a time) times $(2^{6-1}) = 6 * 32 = 192$ different tone circuits. The relation for 2-over-4 coils is $Voc \rightarrow 4V_A \pm 4V_B \pm 2V_C \pm 2V_D \pm 2V_E \pm 2V_F$, giving (6 things taken 2 at a time) times $32 = 480$ different tone circuits. The relation for 3-over-3 coils is $Vo \rightarrow V_A \pm V_B \pm V_C \pm V_D \pm V_E \pm V_F$, giving (6 things taken 3 at a time) times $1/2$ times $32 = 320$ different tone circuits, for a total of $192 + 480 + 320 = 992$ different tone circuits.

The total of all the different tone circuits over all 32 different pole configurations for $J = 2$ to 6 for $K = 6$ is $30 + 240 + 840 + 1440 + 992 = 3542$. Each of the 32 pole configurations share a different set of 301 different tone circuits from a pool of 3542, with $9632 - 3542 = 6090$ duplicates in the set of output selections.

8.2.6 Summary of Results

Fender (US3,290,424, 1966) managed to put 4 sets of double poles under a pick guard, which arguably could have just as well been 8 pickups. Whether or not it would be useful is another matter. For stringed instruments like pianos, where many more pickup coils can be used along the strings, the method of calculating the number of possible humbucking circuits can be easily expanded by the same rules. So for 2, 3, 4, 5, 6, 7, 8, 9, and 10 matched pickup coils, this switching system can produce, respectively, 1, 6, 25, 90, 301, 966, 3025, 9330, and 28,501 humbucking circuits. The natural logs of the number of HB circuits, N_{HB}, are about: 0, 1.79, 3.22, 4.50, 5.70, 6.87, 8.01, 9.14, and 10.26. So the rise in the number of circuits is clearly an exponential function of the number of pickups.

Table 8.4 Numbers of circuits for K pickups taken J at a time in a common connection point switching circuit, for one pole configuration

$J =$	2	3	4	5	6	7	8	9	10	11	12	Totals
K												
2	1											1
3	3	3										6
4	6	12	7									25
5	10	30	35	15								90
6	15	60	105	90	31							301
7	21	105	245	315	217	63						966
8	28	168	490	840	868	504	127					3025
9	36	252	882	1890	2604	2268	1143	255				9330
10	45	360	1470	3780	6510	7560	5715	2550	511			28,501
11	55	495	2310	6930	14,322	20,790	20,955	14,025	5621	1023		86,526
12	66	660	3465	11,880	28,644	49,896	62,865	56,100	33,726	12,276	2047	261,625

Table 8.4 shows these calculations for this kind of circuit extended to K pickups taken J at a time, where $K = 2$–12 and $J = 2$–12. The first thing that becomes apparent is that for J pickups taken J at a time, the number of circuits is $2^{(J-1)} - 1$. Equation 8.5 shows the full equation. This determines the upper limit of switched circuits of this type.

$$\#\text{Circuits for } K = J : 2^{J-1} - 1, \quad J \geq 2$$

$$\#\text{Circuits for } K > J : \left(2^{J-1} - 1\right)\binom{K}{J}, \quad J \geq 2 \tag{8.5}$$

$$\text{Total}\,\#\text{Circuits for } K \geq 3, J \geq 2 : \left(2^{J-1} - 1\right) * \left(1 + \sum_{J=2}^{K-1}\binom{K}{J}\right)$$

Table 8.5 Total number of tone circuits for K pickups taken J at a time in a common connection point switching circuit, over all 2^{K-1} pole configurations

$J =$	2	3	4	5	6	Totals
$2^{J-1} =$	2	4	8	16	32	
K						
2	2					2
3	6	12				18
4	12	48	56			116
5	20	120	280	240		660
6	30	240	840	1440	992	3542

Table 8.5, as stated shows the total number of different no-load tone circuits over all the 2^{K-1} pole configurations, with totals for all the circuits of $J \leq K$. Note in each case that they are larger than the numbers in Table 8.4. And while they can be much smaller than the numbers in Table 7.11 in Sect. 7.3.16, they are still better than 3-way, 5-way or 13-way switching systems, increasing nearly exponentially with the number of available pickups.

References

Lesti, A. (1936). Electric translating device for musical instruments, US Patent 2,026,841, 7 Jan 1936. Retrieved from https://patents.google.com/patent/US2026841A/

Baker, D. L. (2017a). Humbucking switching arrangements and methods for stringed instrument pickups, US Patent Application 15/616,396, filed 7 June 2017, published as US-2018-0357993-A1, 13 Dec 2018, granted as Patent US10,217,450, 26 Feb 2019. Retrieved from https://www.researchgate.net/publication/335727402_Humbucking_switching_arrangements_and_methods_for_stringed_instrument_pickups_-_NPPA_15616396

Baker, D. L. (2018b). Means and methods for switching odd and even numbers of matched pickups to produce all humbucking tones, US Patent Application 16/139,027, 22 Sep 2018, published as US-2019-0057678-A1, Feb 21, 2019, granted as U.S. Patent 10,380,986, 08/13/2019. Retrieved from https://www.researchgate.net/publication/335728060_NPPA-16-139027-odd-even-HB-pu-ckts-2018-06-22

Fender, C. L. (1966). Electric guitar incorporating improved electromagnetic pickup assembly, and improved circuit means, US Patent 3,290,424, 6 Dec 1966. Retrieved from https://patents.google.com/patent/US3290424A/

Chapter 9
A Common-Point Connection Experiment with Two Mini-Humbuckers

9.1 An Experiment with Two Mini-Humbuckers

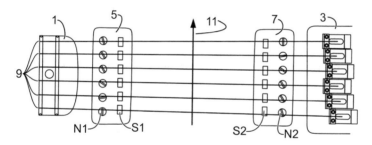

Fig. 9.1 Shows the neck (**1**) to bridge (**3**) region on an electric guitar with two generic Hofner-style mini-humbuckers (**5** and **7**) installed under the strings (**9**). The neck pickup (**5**) and the bridge pickup (**7**) have one set of adjustable screw poles for the N-up poles, (N1) and (N2), with hemispherical heads that extend above the pickup cover, and one set of rectangular S-up poles, (S1) and (S2), that sit flush with the cover. Since both pickups are the same model number, the coils are reasonably matched in response to external hum. The strings are tuned to the standard E-A-D-G-B-E, and were strummed midway between the pickups at (**11**). For reference, Table 9.1 shows the string fundamental and harmonic frequencies

Table 9.1 String fundamental frequencies and harmonics for standard EADGBE tuning (Hz)

String	Fund	2nd Harm	3rd Harm	4th Harm	5th Harm	6th Harm	7th Harm	8th Harm
E	82.4	164.8	247.2	329.6	412	494.4	576.8	659.2
A	110.0	220.0	330.0	440.0	550	660	770	880
D	146.8	293.6	440.4	587.2	734	880.8	1027.6	1174.4
G	196.0	392.0	588.0	784.0	980	1176	1372	1568
B	246.9	493.8	740.7	987.6	1234.5	1481.4	1728.3	1975.2
E	329.6	659.2	988.8	1318.4	1648	1977.6	2307.2	2636.8

© Springer Nature Switzerland AG 2020
D. L. Baker, *Sensor Circuits and Switching for Stringed Instruments*,
https://doi.org/10.1007/978-3-030-23124-8_9

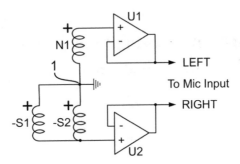

Fig. 9.2 Shows a humbucking triple using coils N1, S1, and S2, with a common connection point (**1**), and two voltage follower preamplifiers, U1 and U2. The output of U1 represents the high output terminal, going to the left microphone channel of the mic input of a desktop PC. The output of U2 represents the low output terminal, going to the right microphone channel

A shareware program, Simple Audio Spectrum Analyzer v3.9, © W.A. Sterr 2001–2006, SpecAn_3v97c.exe, digitized the signal and produced a magnitude-only FFT spectrum for the mic signal in Fig. 9.2, Vo/2 = (Left − Right)/2. It took Hann (raised cosine) windows of 4096 values at a rate of 8000 samples per second, providing a frequency resolution of about 2 Hz, over a range from 0 to 3998 Hz. It averaged all the windows together to produce a discrete FFT spectrum, measured as dB full scale (dBFS) versus frequency, exported into a ∗.CSV text file and imported into MS Excel for processing.

Equation 9.1 shows the equations used to process this FFT data in a spreadsheet. There are 2048 magnitude values in the dBFS scale for frequency bins from 0 to 3998 Hz, with a resolution of about 1.95 Hz. These are converted to linear values, $\lin V_n(f_n)$, which are summed to obtain the relative signal amplitude. Dividing each magnitude by the total provides a probability density function, $P_V(f_n)$, which sums to 1. Multiplying and summing over the product of all the bin frequencies and the density function values gives the mean frequency in Hz. The second and third moments of the FFT spectrum are the bin frequency minus the mean, raised to the second and third powers, times the density function. For the purpose of simply maintaining smaller and more comparative numbers to consider, the second and third roots of the second and third moments have units of Hz.

$$\lin V_n(f_n) = 10^{\mathrm{dBFS}_n/20}, \quad 1 \le n \le 2048$$

$$\text{Relative signal amplitude} = \sum_{n=1}^{2048} \lin V_n$$

$$P_V(f_n) = \frac{\lin V_n}{\displaystyle\sum_{n=1}^{2048} \lin V_n}$$

$$\text{Mean} \cdot f = \sum_{n=1}^{2048} f_n * P_V(f_n)$$

$$\text{2nd} \cdot \text{moment} \cdot f = \sum_{n=1}^{2048} (f_n - \text{mean} \cdot f)^2 * P_V(f_n)$$

$$\text{3rd} \cdot \text{moment} \cdot f = \sum_{n=1}^{2048} (f_n - \text{mean} \cdot f)^3 * P_V(f_n)$$

(9.1)

Table 9.2 shows the results of this experiment for the 25 HB circuits from the 4 coils in Fig. 9.1. The designation "o" between the pole designations means "over," as in N1oS1, means that the N1 signal is connected to the Left or high output in Fig. 9.2, and $-S1$ signal is connected to the Right or low output in, providing the measured output signal, Vo/2 = (Left − Right)/2 = $(V_{N1} + V_{S1})/2$. Likewise, S1oN1N2S2 indicates that the $-V_{S1}$ signal is connected to the high output, and the parallel connection of the signals V_{N1}, V_{N2}, and $-V_{S2}$ signals are connected to the low output, providing a measured signal of Vo/2 = (Left − Right)/2 = $(-V_{S1} + (V_{S2} - V_{N1} - V_{N2})/3)/2$. Hereafter, when the measured results are converted from dBFS to linear, the linear results are multiplied by 2, and the "/2" is dropped. The relative amplitudes in Table 9.2 have been multiplied by 2 after calculation to get the correct value of Vo.

Table 9.2 HB circuits from 4 coils, w/ relative signal amplitudes and root moments

Coils	Relative signal amplitude	Moments (Hz)		
		1st	Root-2nd	Root-3rd
N1oS1	2.83	636.1	684.2	1224.3
N1oN2	1.15	843.0	752.3	1387.5
N1oS2	2.05	713.5	722.7	1295.3
S1oN2	2.31	770.5	740.1	1337.7
S1oS2	0.88	835.0	752.8	1380.1
N2oS2	2.59	907.5	771.0	1440.7
N1oS1N2	0.78	933.1	794.6	1474.9
N1oN2S2	0.23	1201.1	873.1	1724.2
N1oS1S2	2.59	669.8	717.1	1275.0
S1oN1S2	1.91	655.1	704.8	1252.0
S1oN2S2	1.33	637.4	687.4	1226.4
S1oN1N2	2.23	672.2	704.6	1259.0
N2oN1S1	0.31	849.3	824.7	1468.2
N2oN1S2	0.74	712.6	718.1	1288.1
N2oS1S2	2.18	792.8	752.8	1363.7
S2oN1S1	0.36	837.2	822.7	1454.7
S2oN2S1	1.30	683.4	714.9	1274.3
S2oN1N2	2.64	792.9	754.2	1362.3
N1oS1N2S2	0.40	633.2	708.9	1247.4
S1oN1N2S2	0.63	632.9	699.4	1235.3
N2oN1S1S2	0.26	854.7	756.4	1398.3
S2oN1S1N2	0.49	827.6	783.5	1413.4
N1N2oS1S2	2.55	741.4	743.2	1329.1
N1S2oS1N2	1.02	837.0	750.1	1379.9
N1S1oN2S2	0.25	1006.8	868.2	1598.4

The notation N1oS1 means the N1 pickup coil over the S1 pickup coil in Fig. 9.2, with the string signal for the N1 coil going to the Left Mic Input, and the string signal for the S1 pickup coil going to the Right Mic Input. There are 6 humbucking pairs, 12 humbucking triples, and 7 humbucking quads, conforming to Sect. 8.2.6, Table 8.4, the line for $K = 4$

Table 9.3 The 25 results from Table 9.2, ordered by mean frequency (1st moment) from low to high, from 632.9 to 1201.1 Hz

Coils	Relative linear signal amplitude	Moments (Hz)		
		1st	Root-2nd	Root-3rd
S1oN1N2S2	0.63	632.9	699.4	1235.3
N1oS1N2S2	0.40	633.2	708.9	1247.4
N1oS1	2.83	636.1	684.2	1224.3
S1oN2S2	1.33	637.4	687.4	1226.4
S1oN1S2	1.91	655.1	704.8	1252.0
N1oS1S2	2.59	669.8	717.1	1275.0
S1oN1N2	2.23	672.2	704.6	1259.0
S2oN2S1	1.30	683.4	714.9	1274.3
N2oN1S2	0.74	712.6	718.1	1288.1
N1oS2	2.05	713.5	722.7	1295.3
N1N2oS1S2	2.55	741.4	743.2	1329.1
S1oN2	2.31	770.5	740.1	1337.7
N2oS1S2	2.18	792.8	752.8	1363.7
S2oN1N2	2.64	792.9	754.2	1362.3
S2oN1S1N2	0.49	827.6	783.5	1413.4
S1oS2	0.88	835.0	752.8	1380.1
N1S2oS1N2	1.02	837.0	750.1	1379.9
S2oN1S1	0.36	837.2	822.7	1454.7
N1oN2	1.15	843.0	752.3	1387.5
N2oN1S1	0.31	849.3	824.7	1468.2
N2oN1S1S2	0.26	854.7	756.4	1398.3
N2oS2	2.59	907.5	771.0	1440.7
N1oS1N2	0.78	933.1	794.6	1474.9
N1S1oN2S2	0.25	1006.8	868.2	1598.4
N1oN2S2	0.23	1201.1	873.1	1724.2

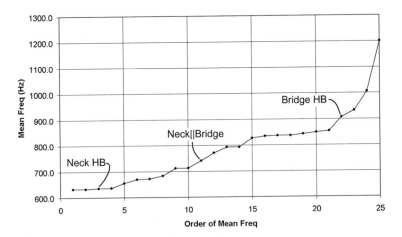

Fig. 9.3 Shows the results of humbucking circuits from Table 9.3 for mean frequency versus frequency order. It highlights the equivalent 3-way switch results, the neck humbucker (Neck HB) at the 3rd spot, 636.1 Hz, the neck and bridge humbuckers in parallel (Neck||Bridge) at the 11th spot, 741.4 Hz, and the bridge humbucker (Bridge HB) at the 22nd spot, 907.5 Hz. It shows a number of frequencies bunched closely together, at 632.9–639.4 Hz, 669.8 and 672.2 Hz, 712.6 and 713.5 Hz, 792.8 and 792.9 Hz, and from 835.0 to 837.2 Hz

Note that the four results above 854.7 Hz have a much steeper curve, and the top three have a lower signal strength, and that the results in general tend to be bunched at the low end, at the presumably warmer tones, and again in the middle-high range between 800 and 900 Hz (Fig. 9.3). Without having done the measurements, one can only speculate that the distribution may have be more even for four matched and evenly spaced pickups, as described in US 9,401,134 (Baker, 2016b).

This suggests that there may be only 17 distinct tones available, a result consistent with a two-humbucker experiment in NPPA 15/616,396 (Baker, 2017a) using a 20-circuit switch. Note also that the relative signal strengths run from 0.23 to 2.83, a factor of 12.3, or about 22 dB. This data will be used to demonstrate a method for ordering tones and choosing switching connections accordingly, with variable gains to equalize signal strengths.

9.2   Humbucker Magnitude Spectra

This author has claimed in other writings[1] that pickups placed close together may have entangled magnetic fields and are too close to generate tonal responses that are significantly different. Simple experiments driving one coil of a humbucker with an audio signal and measuring the output on the other coil show the humbucker to be a very poor transformer. The results obtainable by a shareware program[2] that only delivers magnitude spectra are a bit equivocal. While they show that putting the signals of humbucker coils in contra-phase produce a much lower signal that in-phase results, other combinations in Tables 9.2 and 9.3 where one might expect similar results due to coil placement, such as N1oS2 and S1oN2, still show significant differences.

Better experiments are planned, but will not be ready in time for this book edition. Nevertheless, here are some spectra plots from the experiment above, to be taken with a grain of salt.

---

[1]http://TulsaSoundGuitars.com/tutorials/humbucking-pairs/
[2]http://www.techmind.org/audio/specanaly.html

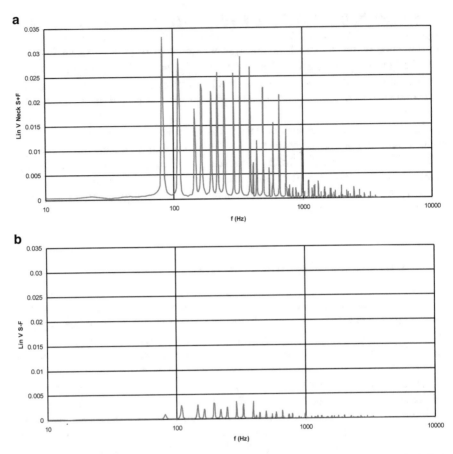

Fig. 9.4 (**a**) Shows the magnitude spectrum of the linear voltage conversion for the full-scale decibel output of the (Left−Right)/2 = ($V_{N1}+V_{S1}$)/2 signal for the setup in Fig. 9.2, using the SpecAn shareware package, with N1 (screw poles on neck HB in Fig. 9.1) in the Left channel and S1 (flat poles on the neck HB in Fig. 9.1) in Right channel. In other words, the magnitude spectrum of the linear full-scale output of N1 over S1. (**b**) The magnitude spectrum of the string signal of N1 in Fig. 9.1 minus the string signal of S1 in Fig. 9.1, ($V_{N1}-V_{S1}$)/2, averaged over 5 strums and converted to lin$V(f)$ in Eq. (9.1)

In Fig. 9.4a, b, SpecAn_3v97c was set to amplitude scale = 100 dBlog; zero weighted; freq scale = log; visualization = Spectrograph w/avg; Sample rate = 8 kHz; FFT size = 4096; FFT Window = Hann (raised cosine); Mic level = 9. The output amplitude of the guitar was set to avoid overdriving the computer microphone input, as determined by the color signal in the software graphical user interface. The middle of the strings (11 in Fig. 9.1) were strummed five times at about one strum per second, then allowed to die off for a few seconds. Note in Fig. 9.2 that Right channel sees the negative string signal of any S-up coil attached it. So the Left − Right output chosen in SpecAn would correspond to N1 + S1 (Fig. 9.4a), and Left + Right would correspond to N1 − S1 (Fig. 9.4b).

This produced frequency magnitude spectra which were exported to CSV text files. The text files were imported as CSV files into MS Excel and averaged together. Then Eq. (9.1) was applied. Figure 9.4a, b shows the lin$V_n(f_n)$ in Eq. (9.1). Figure 9.4a shows the string signals of the neck humbucker coils in-phase, while

Fig. 9.4b shows them contra-phase. Obviously, they produce very similar signals, to be reduced by so much in contra-phase. But not exactly the same, as shown below in the probability density plots of the same signal sums in Fig. 9.5a–c.

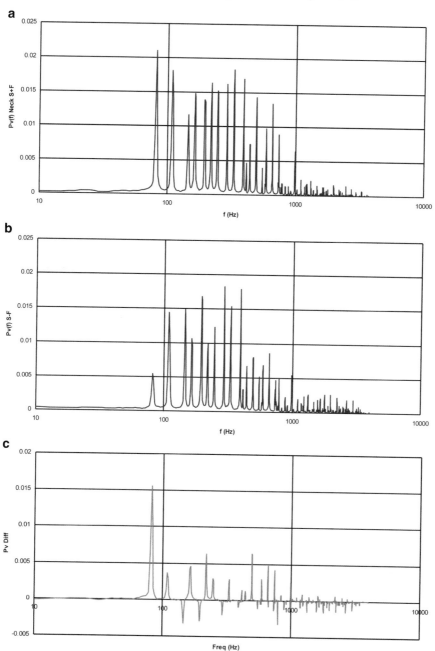

Fig. 9.5 (**a**) The probability density function, $P_V(f)$, from Eq. (9.1), applied to the sum of the string signals of N1 and S1 from Fig. 9.1 versus log frequency (Hz). (**b**) The probability density function, $P_V(f)$, from Eq. (9.1), applied to the difference of the string signals of N1 and S1 from Fig. 9.1 versus log frequency (Hz). (**c**) The difference of the probability density functions in (**a**, **b**)

Remember that the probability density function only shows the distribution of frequencies in the signal in relative amplitude, not the absolute amplitude of signals at those frequencies. Figure 9.5c shows the largest difference between the in-phase and contra-phase signals at 82 about Hz, or low E on the strings. The next two biggest differences stand at 220 Hz (2nd harmonic of A) and 494 Hz (6th harmonic of low E). A real musician might hear the difference, but this author cannot. This author has many more plots of this type from this experiment, but doubts their significance and usefulness, since half the spectral information, the phase plot, is missing.

A complex spectrum, with real and imaginary parts, would be even more useful, but at the time of this writing, January 2019, is not yet in this author's resources.

9.3 A Method of Choosing the Spacing and Switching Order of Tones

This material is covered by US Patent 10,380,986 (Baker, 2019b). The object of the exercise is to offer a much wider range of tones, and to allow the musician to use one control to shift progressively from bright to warm and back, without ever needing to know which pickups are used in what circuit. For that, one needs a way to order the tones.

There is no guarantee at this time that using the mean frequency of the signal from one or more strummed strings, with either open fretting or some chord, will correspond exactly to brightness or warmness of tone, as commonly perceived by a musician's ear. For example, French (2009) noted in a section on psychoacoustics, pp 190–193, that louder tones mask nearby tones. And on pp. 29–36, in a section on human perception of sound, he notes that the sensitivity of human hearing to tones peaks at 1000–2000 Hz. This method of ordering tones needs a simple one-number measure of tone that has not yet been developed and proven. But the mean frequency of six strummed strings is a start, used here as an example until better methods come along.

The mean-frequency numbers used here for illustrating the method come from Eq. (9.1) and Table 9.3, from the dual-humbucker experiment previously disclosed, which also helps to illustrate the method. Ideally, the frequency resolution should be 1 Hz, with a range of from 0 Hz to a top end of at least 4 kHz, but preferably the full range of human hearing, which extends to 20 kHz or more. Preferably, enough sample windows should be taken to cover from the very beginning of a strummed or plucked note or chord through the full sustain of the sound. But it may turn out that other sampling techniques have certain advantages not discussed here.

One should expect that, like the dual-humbucker experiment, some tones will be too close together to count, and the separation of tones with switched pickups circuits will vary considerably, likely with most of the tones bunched together at the warm end. So, for four pickups with 25 different humbucking circuits, there may be only half that number of useful tones. And for 25 different circuits and a six-throw switch, only half of those dozen can be used. For pickups with reversible poles, four pickups have 8 different pole configurations, sharing 25-member sets of 116 potentially unique tones. (The ratio of the numbers [poles times circuits] to the total numbers of tones is always greater than or equal to one.)

Digitally controlled analog switching may have a much wider range of choice than mechanical switches, but the problem of bunched tones still exists. Note that in Table 9.3, the range of mean frequency from 632.9 Hz at the low end to 1201.1 Hz at the high end, for one pole configuration, is barely an octave. Without actual measurements, it is not yet possible to know what other pole configurations will produce. Nor is it yet possible to account for the variations introduced by moving pickups themselves about in space, as disclosed in US 9,401,134 B2 (Baker, 2016b), offering 5 degrees of freedom, vertically and along the strings at each end of a pickup, as well as across the strings.

This method assumes that whatever the measure of tone, it should be divided along bright to warm, or warm to bright, according to virtual frets. In most Western music, adjacent notes differ by a multiplier or divisor of $2^{1/12}$, counting 0–12 frets from an open note to its octave note. Other musical traditions can have three times as many notes per octave. This division of frequencies comes from the way that the human ear is constructed and responds to sound. So it is natural to assume that the most effective and efficient way to chose the separation of tones chosen and ordered from those available is by a constant frequency multiplier from one tone to the next higher tone.

The method disclosed here is fairly simple: (1) chose a measure of tone (mean frequency of six strummed strings from FFT analysis in these examples); (2) cause the musical instrument to emit tones in some standard fashion (strum six strings several times in these examples); (3a) take digital acoustic samples of the signal outputs from each pickup simultaneously (not quite possible in these examples), or alternatively, (3b) take digital acoustic samples from each switched pickup circuit; (4) digitally process the acoustic samples to obtain complex number frequency spectra for each pickup or each pickup circuit (only magnitudes of frequency bins were possible for these examples, leaving out phase information); (5) apply the measure of tone to the individual frequency spectra (Eq. (9.1) and Table 9.3 in these examples); (6) pick the range of tones (from mean frequencies in Table 9.3 in these examples); (7) pick the number of tones to be switched (for example, six tones for a 6-throw switch); (8) calculate the virtual fret steps between switched tones; (9) choose the closest available tones to those steps; and (10) wire or program the mechanical and digital-analog switch to select the circuits that produce those tones.

Since human hearing is very subjective, there's an alternative extension to the method that orders the tones according to the musician's preference. Anytime after step 4, when the samples have been taken and FFT transforms have been stored, the inverse-FFT transform can convert the spectra back into a string of sounds. The sound that comes out will be the average of all the sample windows taken over the entire original length of the notes. So the strike and decay of the sound may be averaged together.

It's the Optometrist approach, and requires either the use of a micro-controller with a digital-to-analog converter to produce the sounds and ask the musician for decisions, or presentation by a person customizing the guitar. The inverse-FFT characteristic sound of each of two switched circuits plays back to the musician, and the software asks, "Which sound is warmer? Tone A? Or Tone B?" Or, the guitar customizer simply plays the tones on the guitar and asks the same questions. Then the musician picks, and the use of an efficient sorting algorithm, such as a shell sort, determines the order of the tones for switching. Then the entire set is played back in order for confirmation and adjustment.

The following examples include equations and tables to help illustrate the method.

9.3.1 Example 1: Choosing 6 Tones from Table 8.10 Using Mean Frequency for a 6P6T Switch

Suppose that the only switch available is a 6P6T mechanical switch, and we wish to use the entire frequency range in Table 8.10 from 632.9 to 1201.1 Hz. Equation 9.2 shows a simple way to calculate the ratio between frequency steps, r, where the lowest frequency, 632.9 Hz, is multiplied by r five times to get the highest frequency and all the steps in between for a 6-throw switch.

$$
\begin{aligned}
&\text{(a) } 1201.1 = r^5 * 632.9 \\
&\text{(b) } r = \sqrt[5]{1201.1/632.9} = 1.13671\ldots \\
&\text{(c) Note: } r = 2^{\frac{a}{5*12}} = 2^{\frac{a}{60}} \\
&\text{(d) } a = 60\frac{\ln(r)}{\ln(2)} = 11.0917\ldots \text{ fret steps}
\end{aligned} \tag{9.2}
$$

Throw	1	2	3	4	5	6
Freq (Hz)	632.9	719.4	817.8	929.6	1056.6	1201.1

It is usually not possible to use the measured mean frequencies to hit those marks exactly. So one takes the choices that seem best. The first frequency, 632.9 Hz, has a pickup combination, S1 over N1N2S2, a quad circuit. The closest ones to 719.4 Hz are 712.6 at 0.74 relative amplitude and 713.5 at 2.05 amplitude. The best choice is 713.5 Hz, from combination N1 over S2. The 3rd frequency, 817.8 Hz, is 24.9 Hz up from 792.9 and 9.8 Hz down from 827.0 Hz. If signal strength is important, then the lower frequency would be better, but the relative amplitude of the highest frequency output, 1201.1 Hz only has a relative amplitude of 0.23, so S2 over N1S1N2 at 827.0 Hz it is. The closest and only choices for 929.6 and 1056.6 Hz are N1 over S1N2 at 933.1 Hz and N1S1 over N2S2 at 1006.8 Hz, leaving N1 over N2S2 at 1201.1 Hz. Table 9.4 shows the chosen order brightest to warmest tones, according to the mean frequencies of 6 strummed strings.

Table 9.4 Order of tones from 1202 to 633 Hz for a 6P6T switch

Throw	1	2	3	4	5	6
Pickup circuit	N1 N2S2	N1S1 N2S2	N1 S1N2	S2 N1S1N2	N1 S2	S1 N1N2S2
Signal	N1 − N2 + S2	N1 − S1 −N2 + S2	N1 + S2 − N2	−S2 − N1 +S1 − N2	N1 + S2	−S1 − N1 −N2 + S2
Mean freq (Hz)	1201.1	1006.8	933.1	827.6	713.5	632.9
~Fret number	11.1	8.0	6.7	4.6	2.1	0
Relative amplitude	0.23	0.25	0.78	0.49	2.05	0.63

From Table 14, NPPA 16/139,027, US Patent 10,380,986

Compare this to Table 9.5, representing a 3-way switch giving the bridge HB, the neck and bridge HB in parallel, and the neck HB.

Table 9.5 Outputs for a standard 3-way switch. From Table 15, NPPA 16/139,027, US Patent 10,380,986

Throw	1	2	3
Pickup circuit	N2 S2	N1N2 S1S2	N1 S1
Signal	N2 + S2	N1 + N2 + S1 + S2	N1 + S1
Mean freq (Hz)	907.5	741.4	636.1
~Fret number	6.2	2.7	0
Relative amplitude	2.59	2.55	2.83

9.3.2   Example 2: Choosing 6 Tones from Table 9.6 Using Weighted Moments

Suppose it should be determined that a better measure of tones comes from giving a weight of 1 to the mean frequency, ½ to the square root of the 2nd moment, and 1/3 to the 3rd root of the 3rd moment in Table 9.3. The normalized fractions would be 6/11 of the mean frequency, 3/11 of the root 2nd moment and 2/11 of the root 3rd moment, as shown ordered by Weighted moments in Table 9.6.

Table 9.6 Coil circuits ordered by weighted moments, Weighted $= 6 * (\text{1st})/11 + 3 * (\text{root-2nd})/11 + 2 * (\text{root-3rd})/11$

Coils	Signal amplitude	Moments (Hz)			
		1st	Root-2nd	Root-3rd	Weighted
N1oS1	2.83	636.1	684.2	1224.3	756.2
S1oN2S2	1.33	637.4	687.4	1226.4	758.1
S1oN1N2S2	0.63	632.9	699.4	1235.3	760.6
N1oS1N2S2	0.40	633.2	708.9	1247.4	765.5
S1oN1S2	1.91	655.1	704.8	1252.0	777.2
S1oN1N2	2.23	672.2	704.6	1259.0	787.7
N1oS1S2	2.59	669.8	717.1	1275.0	792.7
S2oN2S1	1.30	683.4	714.9	1274.3	799.4
N2oN1S2	0.74	712.6	718.1	1288.1	818.7
N1oS2	2.05	713.5	722.7	1295.3	821.8
N1N2oS1S2	2.55	741.4	743.2	1329.1	848.7
S1oN2	2.31	770.5	740.1	1337.7	865.3
N2oS1S2	2.18	792.8	752.8	1363.7	885.7
S2oN1N2	2.64	792.9	754.2	1362.3	885.8
S1oS2	0.88	835.0	752.8	1380.1	911.7
N1S2oS1N2	1.02	837.0	750.1	1379.9	912.0
N1oN2	1.15	843.0	752.3	1387.5	917.3
S2oN1S1N2	0.49	827.6	783.5	1413.4	922.1
N2oN1S1S2	0.26	854.7	756.4	1398.3	926.7
S2oN1S1	0.36	837.2	822.7	1454.7	945.5
N2oN1S1	0.31	849.3	824.7	1468.2	955.1
N2oS2	2.59	907.5	771.0	1440.7	967.2
N1oS1N2	0.78	933.1	794.6	1474.9	993.8
N1S1oN2S2	0.25	1006.8	868.2	1598.4	1076.5
N1oN2S2	0.23	1201.1	873.1	1724.2	1206.7

From Table 16, NPPA 16/139,027, US Patent 10,380,986

Suppose that the same 6-throw switch will be used, with 756.2 Hz the lowest tone, 1206.7 Hz the highest tone, and 4 in between, all separated by the same frequency multiplier. Equation 9.3 shows the calculations.

$$
\begin{aligned}
&\text{(a) } 1206.7 = r^5 * 756.2 \\
&\text{(b) } r = \sqrt[5]{1206.7/756.2} = 1.097975\ldots \\
&\text{(c) Note: } r = 2^{\frac{a}{5*12}} = 2^{\frac{a}{60}} \\
&\text{(d) } a = 60\frac{\ln(r)}{\ln(2)} = 8.09\ldots \text{fret steps}
\end{aligned} \tag{9.3}
$$

Throw	1	2	3	4	5	6
Freq(Hz)	756.2	830.3	911.6	1001.0	1099.0	1206.7

For 830.3 Hz, 821.8 is 8.5 Hz below and 848.7 is 18.4 above, leaving 821.8 Hz the closest. For 911.6 Hz, 911.7 is closest. For 1001.0 Hz, 993.8 Hz is closest, leaving 1076.5 for 1099.0 and 1206.7 Hz. Table 8.16 shows the results of these choices. Because of the dearth of choices at the high end, only the choices for throws 4 and 6 have changed from Table 9.7.

Table 9.7   Order of 6 tones from 1207 to 756 Hz for Weighted moments

Throw	1	2	3	4	5	6
Pickup circuit	N1 N2S2	N1S1 N2S2	N1 S1N2	S1 S2	N1 S2	N1 S1
Signal	N1 − N2 + S2	N1 − S1 − N2 + S2	N1 + S1 − N2	−S1 + S2	N1 + S2	N1 + S1
Mean freq (Hz)	1206.7	1076.5	993.8	911.7	821.8	756.2
~Fret number	8.1	6.1	4.7	3.2	1.4	0
Relative amplitude	0.23	0.25	0.78	0.88	2.05	2.83

From Table 17, NPPA 16/139,027, US Patent 10,380,986

9.3.3   Example 3: Steps of 1/2 Fret or More from Table 9.3 Using Mean Frequency

Suppose that we wish to remove the near-duplicate tones by specifying that the difference in virtual fret step between tones be 0.5 fret or greater, or a frequency multiplier of $2^{1/24}$, from Table 9.3. Obviously, not all of those slots will be filled, and some closer choice may be sacrificed for another with a larger signal. Table 9.8 shows the first-cut list, choosing 12 out of 25 circuits, with approximate fret steps between mean-frequency choices ranging from 0.5 to 3.1. The first column starts with the first choice, 632.9 Hz, with the value for the 1/2-fret step up in the second column. The next value in the first column is taken from that, either 0.5 fret or more, and so on, except that 933.1 Hz is chosen instead of 934.1 Hz because it is so close. The signal for 792.9 Hz was chosen over 792.8 Hz because it had a stronger signal. The 3rd column shows the relative number of frets from 632.9 Hz; the 4th shows the relative measured amplitude of the signal derived from 6 strummed strings; and the 5th shows the coil connections, with the "+" output shown over the "−" output. The 6th column shows the amplifier gain for each switching combination required to equalize all the signals to the amplitude of the strongest signal, 792.9 Hz for S2 over N1N2. They range from 1.0 to 11.47.

Table 9.8 Half-fret or more steps from Table 9.3

Choice	1/2-Fret up	Fret step	Relative amplitude	Coils	Required gain
632.9	651.4	0.0	0.633	S1 N1N2S2	4.17
655.1	674.3	0.6	1.907	S1 N1S2	1.38
683.4	703.4	1.3	1.297	S2 N2S1	2.03
712.6	733.5	2.1	0.745	N2 N1S2	3.54
741.4	763.1	2.7	2.548	N1N2 S1S2	1.03
792.9	816.1	3.9	2.637	S2 N1N2	1.00
827.6	851.9	4.6	0.489	S2 N1S1N2	5.40
854.7	879.7	5.2	0.261	N2 N1S1S2	10.10
907.5	934.1	6.2	2.588	N2 S2	1.02
933.1	960.4	6.7	0.775	N1 S1N2	3.40
1006.8	1036.3	8.0	0.252	N1S1 N2S2	10.46
1201.1	1236.3	11.1	0.230	N1 N2S2	11.47

From Table 18, NPPA 16/139,027, US Patent 10,380,986

9.3.4 Example 4: Steps of 1/2 Fret or More from Table 9.6 Using Weighted Moments

Table 9.9 shows the same method used for Table 9.8, using weighted moments in Table 9.6, i.e., [6 * (mean-freq)/11 + 3 * (root-2nd)/11 + 2 * (root-3rd)/11] (Hz). In this table, 967.2 Hz with a 0.4 fret step is used because there was nothing else closer, and it allowed 12 tones instead of just 11. This gives a range of fret steps between weighted moments of 0.4–2.0. Under the criterion of 0.5 fret step or more, it could be discarded, leaving 11 tones, and a range of fret steps of 0.5–2.0. The range of gains required to equalize amplitudes goes from 1.0 to 12.32.999

Table 9.9 Half-fret or more steps from Table 9.6 using weighted moments

Choice	1/2 Fret up	Fret step	Relative amplitude	Coils	Required gain
756.2	778.3	0.0	2.83	N1 S1	1.00
777.2	800.0	0.5	1.91	S1 N1S2	1.49
799.4	822.8	1.0	1.30	S2 N2S1	2.18
821.8	845.9	1.4	2.05	N1 S2	1.38
848.7	873.6	2.0	2.55	N1N2 S1S2	1.11
885.8	911.8	2.7	2.64	S2 N1N2	1.07
911.7	938.4	3.2	0.88	S1 S2	3.22
945.5	973.2	3.9	0.36	S2 N1S1	7.86
967.2	995.6	4.3	2.59	N2 S2	1.09
993.8	1023.0	4.7	0.78	N1 S1N2	3.65
1076.5	1108.1	6.1	0.25	N1S1 N2S2	11.24
1206.7		8.1	0.23	N1 N2S2	12.32

From Table 19, NPPA 16/139,027, US Patent 10,380,986

References

Baker, D. L. (2016b). Acoustic-electric stringed instrument with improved body, electric pickup placement, pickup switching and electronic circuit, US Patent 9,401,134, 26 July 2016. Retrieved from https://patents.google.com/patent/US9401134B2/

Baker, D. L. (2017a). Humbucking switching arrangements and methods for stringed instrument pickups, US Patent Application 15/616,396, filed 7 June 2017, published as US-2018-0357993-A1, 13 Dec 2018, granted as Patent US10,217,450, 26 Feb 2019. Retrieved from https://www.researchgate.net/publication/335727402_Humbucking_switching_arrangements_and_methods_for_stringed_instrument_pickups_-_NPPA_15616396

Baker, D. L. (2018b). Means and methods for switching odd and even numbers of matched pickups to produce all humbucking tones, US Patent Application 16/139,027, 22 Sep 2018, published as US-2019-0057678-A1, Feb 21, 2019, granted as U.S. Patent 10,380,986, 08/13/2019. Retrieved from https://www.researchgate.net/publication/335728060_NPPA-16-139027-odd-even-HB-pu-ckts-2018-06-22

Baker, D. L. (2019b). Means and methods for switching odd and even numbers of matched pickups to produce all humbucking tones, US Patent Application 16/139,027, 22 Sep 2018, published as US 2019/0057678 A1, 21 Feb 2019, granted as U.S. Patent 10,380,986, 08/13/2019. Retrieved from https://patents.google.com/patent/US10380986B2/

French, R. M. (2009). *Engineering the guitar: Theory and practice*. New York: Springer. Retrieved from https://www.springer.com/us/book/9780387743684

Chapter 10
Switching Systems for Common-Point Connection Pickup Circuits

10.1 Mechanical Switching Systems for Common-Point Connection Circuits

With a common-point connection circuit, a guitar with just two matched single-coil pickups can provide only one no-load humbucking tone circuit for each of the two possible magnetic pole configurations. Three matched pickups can provide 6 HB tone circuits for each of 4 possible pole configurations, and 18 over all the pole configurations. It is possible to set the order of tones with a plug-in connection patch board for each of the 4 pole configurations.

On the other hand, four matched pickups can provide 25 HB tone circuits for each of 8 pole configurations, and 116 over all the pole configurations. So while electrome-chanical switches can be effective for circuits of 2 and 3 pickups, using them for 4 pickups not only wastes a lot of possibilities, it makes it virtually impossible for the switching system to order tones from bright to warm over all. Digitally controlled analog switching systems are more effective for 4, 5, or 6 pickups (and more for pianos).

Table 10.1 Moments (Hz) for humbucking pairs of matched pickup coils from Table 9.2

Coils	Relative signal amplitude	Moments (Hz)		
		1st	Root-2nd	Root-3rd
N1oS1	2.83	636.1	684.2	1224.3
N1oS2	2.05	713.5	722.7	1295.3
S1oN2	2.31	770.5	740.1	1337.7
S1oS2	0.88	835.0	752.8	1380.1
N1oN2	1.15	843.0	752.3	1387.5
N2oS2	2.59	907.5	771.0	1440.7

© Springer Nature Switzerland AG 2020
D. L. Baker, *Sensor Circuits and Switching for Stringed Instruments*,
https://doi.org/10.1007/978-3-030-23124-8_10

10.1.1   Embodiment 1: 3 Matched Pickups with a 4P6T Switch

Consider the humbucking pairs in Table 10.1 from Fig. 9.1 and Table 9.2. The positions of N1 and S1 under the strings are very close together, as are the positions of N2 and S2. So the results of N1 over S2 and S1 over N2 tend to bunch together as do the results of S1 over S2 and N1 over N2. The change between the mean frequency of 636.1–907.5 Hz, which is effectively the change in position of an in-phase pair under the strings, is only about 6.15 virtual frets. The change in virtual frets from 713.5 to 843.0 Hz, which approximates a change in pole configuration, is just under 3 frets. The range of amplitudes is 0.88–2.83, or 1–3.22. So if the amplitudes are equalized to the largest string signal, a gain of up to 3.22 may be needed.

For a 2-coil system with common-point connections, there is no need to switch within one pole configuration. If the magnet of one of the coils is reversible, the 2 different outputs can be handled by a 1P2T (1-pole, 2-throw) switch, to set the gain resistor for a preamp so that the two output signals have equal amplitude. Take the approximating example of N1 over S2 and N1 over N2, with an amplitude ratio of 1.78 to 1. Tone can be handled by either a peak resonant capacitor and pot on the input of the preamp, or an RC-rolloff filter, comprised of a capacitor and pot on the output of the preamp, or both. An active preamp doesn't have to be used, if the relative amplitude differences are tolerable. But a differential preamp is recommended, if only to take advantage of rejecting common-mode noise.

For a 3-coil system, there are only 3 coil terminals to switch over 6 selection choices per pole configuration, as in Tables 8.1 and 8.2. A common 4P6T rotary switch will accommodate the 3 coil terminals, with the 4th pole being used either for a gain resistor to equalize signal outputs, or for a tone capacitor to equalize the low-

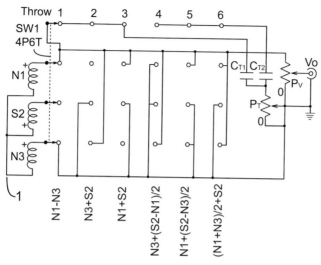

Fig. 10.1 From Fig. 7 in U.S. Patent Application 16/139,027 (Baker, 2018b, US10,380,986), shows a common-connection point circuit with N-up matched pickups N1 (neck) and N3 (bridge) and S-up matched pickup S2 (middle) with their negative hum phase terminals connected to the common point (1), which is floating. Three poles of a 4P6T switch (SW1) connect the remaining terminals of the pickups to the output high terminal and the grounded low terminal so as to produce the connections and signals N1 − N3, N3 + S2, N1 + S2, N3 + (S2 − N1)/2, N1 + (S2 − N3)/2 and (N1 + N3)/2 + S2. The 4th pole connects to the tone capacitor C_{T1} for humbucking pairs and C_{T2} for humbucking triples, with the tone pot P_T. The high circuit output point connects to the volume pot, P_V, which connects to the single-ended output through its wiper terminal

pass frequency roll-off between humbucking pairs and triples, with different lumped impedances. The following figures show various examples of this kind of system, taken from Baker (2018b).

In Fig. 10.1, as is common, dots at line crossings show connections and crossing without dots are pass-overs, as with the lines above C_{T1} and C_{T2}. If each of the matched coils have inductance, L_C, then the first three throws have circuit with a lumped inductance of $2 * L_C$, and the last three have a lumped inductance of $3 * L_C/2$. Tone capacitors C_{T1} and C_{T2} can be used to maintain the equal effect of the tone pot, P_T, on tone. Since resonance frequency is a function of the product of inductance and capacitance, the products $2 * L_C * C_{T1}$ and $3 * L_C * C_{T2}/2$ must be equal to achieve similar tone results, implying that $C_{T1} = 3 * C_{T2}/4$. Both the tone circuit and the volume pot, P_V, lie across the output of the switching circuit. The wiper of the volume pot is connected to the output, Vo.

This is not the only possible selection of matched coils. They could all be either all S-up or all N-up. In which case, all the outputs would be humbucking but out-of-phase, or contra-phase. Without amplification and signal equalization, the output signals would be much weaker, but much brighter. A selection of matched coils that has only one S-up, as shown here, and a selection that has only one N-up will produce the same tones if the opposite poles from each set occupy the same positions under the strings. In other words, N–S–N is the same as S–N–S. In the case of N–S–N, the physical positioning of the S-up pole under the strings will also determine tone, with different sets of tones from S–N–N and N–N–S.

If the pickup magnetic poles are reversed to change the tonal character of the guitar, each pole change will affect both the frequency and order of tones. The order of tones for the switch wiring for one set of poles likely will not hold for another. So there must be some way to change the wiring of the switching along with changing the poles to at least keep an order of tone monotonic from warm to

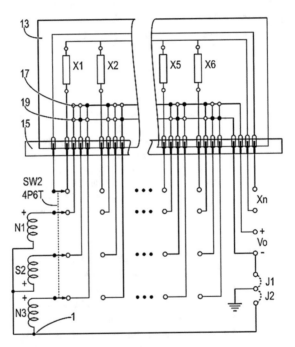

Fig. 10.2 From Fig. 8, USPatApp 16/137/029 (adapted from Fig. 30 in US 9,401,134, Baker 2016b), showing the pickup connections in Fig. 10.1 moved to a plug/patch board (13), plugging into a socket (15), with the high circuit outputs connected to the Vo+ line (17) and the low circuit outputs connected to the Vo– line (19), with switched adjustment components, X1–X6, connected to the terminals Xn. Jumpers J1 and J2 select whether the ground is connected to Vo– (single-ended output) or to the common-connection point (1) (differential output)

bright. US 9,401,134 (Baker, 2016b) disclosed such a device in Fig. 30, a plug-in board with cross-points to be soldered with through-hole jumpers to set the switch connections.

Figure 10.2 shows such a device for this example. One S-up pickup, S1, and two N-up pickups, N1 and N2, have their common connection point (1) connected to jumper, J2, and their other terminals connected to the 4P6T switch, SW2. All of the switch throw inter-connections are made off the switch, on a plug board (13), connected to the throws and one pole by a plug connector (15). The plug connector is shown as a fingerboard connector, but can be anything that fulfills the same function. The figure is split to show throws 1, 2, 5, and 6, but not 3 and 4. One pole and the six related throws of the switch connect to components X1–X6 on the plug board, which can be any kind of printed circuit board, hard or flexible, or anything else that fulfills the function. Components X1–X6 and the associated pole of SW2 are connected off the board to the output, Xn. These components can be resistors for gain control, or capacitors for tone control, or some other function.

The other three times 6 throws, connect through a line of cross-point interconnects (17) to the high output terminal, Vo+, and through another line of interconnects (19) to the low output terminal, Vo−. The vertical circuit lines over the interconnects are on one side of the board and the horizontal lines on the other, so that they do not connect, except through the interconnects. The interconnects can be either non-plated-through holes for soldered through jumpers, or standard computer board jumpers, or some other type that fulfills the function. The white dots show no connection, and the black dots show interconnections. The interconnections shown produce output voltages of Vo = $V_{N3} + V_{S2}$, Vo = $V_{N1} − V_{N3}$, Vo = $(V_{N1} + V_{N3})/2 + V_{S2}$, and Vo = $V_{N1} + (V_{S2} − V_{N3})/2$, for throws 1, 2, 5, and 6, respectively. Any combination and order of humbucking pairs and triples, including duplicates, is possible.

At the output, only one of jumpers J1 and J2 may be connected. If J1 is connected, then the lower terminal of Vo− is grounded, and the output is single-ended, as are most electric guitar circuits. If J2 is connected, then the common pickup connection point is grounded and the output, Vo, is differential. A differential output requires either that a differential amplifier convert it to single-ended, or that the output jack of the electric guitar is stereo, and feeds through 2-conductor shielded cable to a guitar amp with a differential input. A single-ended output has the advantage of using circuits and connections already common to electric guitars. A differential output has the advantage of suppressing common-mode electrical noise from external sources, possibly such as fluorescent lights, which put out much higher frequencies of noise than 60 Hz motors.

Figure 10.2 can be adapted to any electromechanical pickup switching system. Baker (2018a) shows how there can be $2^{J−1}$ pole configurations for J number of matched single-coil pickups with reversible poles, or 4 pole configurations for 3 pickups, each having 6 possible pickup circuits, and 8 configurations for 4 pickups, each having 25 possible pickup circuits, and 16 configurations for 5 pickups, each having 116 possible pickup circuits. This switching system requires a pole for each pickup, and currently the most practical and affordable switches have six poles or less, and six throws or less. For example, with 3 pickups, a 4P6T switch can have one pole dedicated to a set of adjustment components, resistive or capacitive or something else, and a 5P6T switch can have two poles dedicated to adjustment components, say resistive for gain control with active circuits and capacitive for tone control.

10.1.2   *Embodiment 2: 3 Matched Pickups with a Single-Ended Preamp and Signal Volume Compensation*

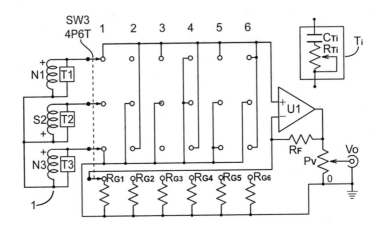

Fig. 10.3 From Fig. 10, USPatApp 16/139,027, similar to Fig. 10.1 above, but with the 4th pole of 4P6T switch, SW3, selecting the gain resistors (or pots) R_{G1} to R_{G6}, for the single-ended preamp circuit of U1, R_F and R_{Gi}, with gain $= 1 + R_F/R_{Gi}$. Tone/peaking circuits, T1, T2 and T3, comprised of tone capacitors $C_{T1} - C_{T3}$ and tone pots $R_{T1} - R_{T3}$, have been added to each pickup

Figure 10.3 shows this embodiment, with pickups N1 and N2 N-up, and S1 S-up. The 4P6T switch, SW3, uses 3 poles in a switching system with a common connection point (1) to connect the pickups by throws: (1) N1 over N2, or N1 − N2; (2) N2 over S1, or S1 + N2; (3) N1 over S1, or N1 + S1, (4) N2 over N1 and S1, or N2 + (S1 − N1)/2; (5) N1 over S1 and N2, or N1 + (S1 − N2)/2; and (6) N1 and N2 over S1, or (N1 + N2)/2 + S1, where the pickup designations also stand in for signal voltages. The circuit uses the 4th pole to switch gain resistors, R_{G1} to R_{G6} into a circuit using operational amplifier U1 as a single-ended preamp with a feedback resistor, R_F. Equation 10.1 shows how the gains and R_G values are calculated. The output of U1 feeds a volume pot, P_V, which goes to the output jack, Vo.

$$Gi = \frac{Vo}{Vs} = \frac{R_{Gi} + R_F}{R_{Gi}} = 1 + \frac{R_F}{R_{Gi}}, \quad R_{Gi} = \frac{R_F}{Gi - 1}$$

$$Gi\# = \frac{Vs\,max}{Vsi}, \quad g = 1 + \frac{R_F}{R_{Gimax}}, \quad Gi = g * Gi\# \qquad (10.1)$$

$$R_{Gi} = \frac{R_F}{Gi - 1}$$

Table 10.2 Example gain resistors for Embodiment 3, Fig. 10, with $R_F = 47k$ and $R_{G1} = 2.2M$

Throw	1	2	3	4	5	6
Relative amplitude	3.161	2.051	2.311	1.148	2.519	0.252
$Gi\# =$	1	1.54	1.37	2.75	1.25	12.54
$Gi = g * Gi\# =$	1.021	1.57	1.40	2.81	1.28	12.81
R_{Gi} (kΩ)	2200	82	128	26	170	4

The tone controls are T1, T2, and T3 across each pickup, each comprised of a tone pot, R_{Ti}, and tone capacitor, C_{Ti}. Note that the lower terminal of the switching system is grounded to the output, so the common connection point cannot be. These tone controls will each react with the respective coils as a resonant peak, high-frequency roll-off circuit. Many single-coil pickups have a self-resonant frequency of 10 kHz or higher. Adding a capacitor alone across a single-coil pickup will cause a resonant peak at a lower frequency, which attenuates frequencies higher than the resonant peak. Adding a tone pot in series with such a capacitor can flatten the resonant peak to a high-frequency roll-off over the range of the pot. Doing this for each pickup individually adds a much wider range of flexibility in controlling the tone of the instrument. One might use small 10–20 turn pots, each accessible with a small or jewelers screwdriver though small holes in the pick guard.

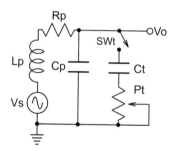

Fig. 10.4 Shows a single-coil pickup with a string signal, Vs, a coil inductance, Lp, a coil resistance, Rp, a coil capacitance, Cp, and an optional (dotted line) tone circuit, comprised of a tone switch, SWt, capacitor, Ct, and a tone pot, Pt. The output is Vo

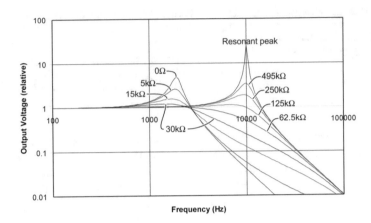

Fig. 10.5 Shows the AC transfer function from Vs to Vo of the circuit in Fig. 10.4, when Lp = 2 H, Rp = 5 kΩ, Cp = 126 pF, Ct = 3.3 nF, and the resistance between Ct and ground is Pt = 495k, 250k, 125k, 62.5k, 30k, 15k, 5k and 0 Ω, when the tone switch, SWt, is closed. The Resonant Peak is the natural 10 kHz peak in the pickup transfer function when SWt is open

Figures 10.4 and 10.5 show the circuit and response curves for a single-coil pickup in Fig. 10.3 with an optional tone circuit, Ti, comprised of a capacitor Ct, and a tone pot, Pt. The curve labeled "resonant peak" is the natural 10 kHz peak response of an imaginary pickup with Lp = 2 H, Rp = 5 kΩ and Cp = 126 pF. If the tone pot, Pt had an open circuit with infinite resistance, the resonant peak curve would be the result. With no tone circuit, the output peaks to about 25 times the input at 10 kHz, and drops to ½ the input at about 17.4 kHz. The rest of the curves in Fig. 10.5 show how the transfer curve changes as the resistance of Pt between Ct and ground decreases from about 500k to 0. In the high-frequency roll-off of each curve, the phase of the output tends to reverse to contra-phase, which may cause some interesting effects in tone. But the output drops rapidly with frequency in the roll-off, quickly limiting the effects in the higher frequencies.

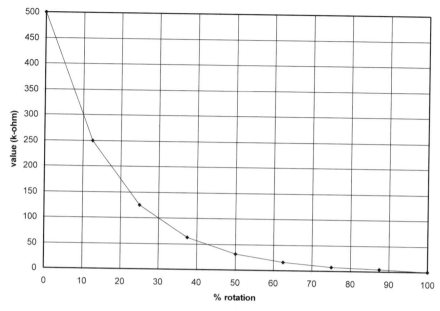

Fig. 10.6 Potentiometer function, a resistance plot for Rt in Fig. 10.4 versus % rotation of shaft. Approximates R (kΩ) $= 500 * \exp(-0.05545 * (\%\text{rotation}))$, to obtain the spread of effects shown in Fig. 10.5

At 495k and 250k, the peak diminishes. At 125k, the curve is more like a low-pass filter where the output signal is ½ the input at about 15 kHz. At 62.5k, the output signal is about ½ the input at about 9550 Hz. At 30k, it the output signal peaks at about 1.2 times the input at about 1445 Hz, and drops to ½ at about 5010 Hz. At 15k, it peaks to about 1.55 times the input at about 1660 Hz, and drops to ½ the input at about 3700 Hz. At 5k, it peaks to about 2.6 times the input at about 1820 Hz, and drops to ½ the input at about 3310 Hz. At zero resistance, it peaks to about 4.9 times the input at about 1900 Hz, and drops to ½ the input at about 3310 Hz (Fig. 10.6).

Since we have no experimental data for a 3-coil guitar, let the imaginary relative signal amplitudes before amplification in Table 10.2 stand in for the sake of argument and example. We conveniently choose the maximum relative signal strength of 3.161 as the first gain, and we wish to adjust the other gains to bring all the other signals up to that level at the output, Vo. Dividing that relative amplitude by all the others, give the relative gain, $Gi\#$, for each signal that we need to approach. But if we pick a feedback resistor, $R_F = 47$k, and a minimum gain resistor $R_{G1} = 2200$k, or 2.2M, then the first gain will be 1.024 instead of 1. We have to multiply this number times all the gains to get the real gains, then calculate R_{Gi}. Equation 10.1 and Table 10.2 show these calculations.

Only a few of the R_{Gi} values are close to standard resistor values. Given that and the differences between human perception and electronic measurements, it would be better to use small, square multi-turn potentiometers for the other R_{Gi}. And if any of the pickup poles are to be reversed, it would be better to use a connection plug board, like that in Fig. 10.2, with the pots mounted in place of the components, Xi.

10.1.3   Embodiment 3: 3 Matched Pickups with a Differential Preamp and Signal Volume Compensation

See Fig. 10.7.

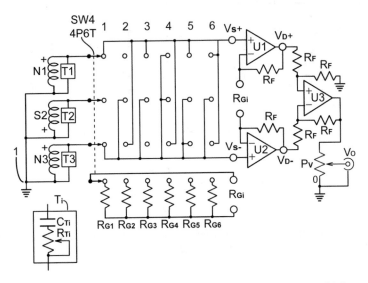

Fig. 10.7 From Fig. 11, NPPA 16/139,027 (Baker, 2018b), showing 2 N-up and 1 S-up matched single-coil pickups, N1, S2, and N3 in positions 1, 2, and 3 under the strings, connected to a grounded common point (1) at hum terminals of the same phase, with "+" signs showing the string vibration signal phase; Tone circuits T1, T2, and T3 across the respective pickups; a 4P6T switch, SW4, with 3 poles switching the pickups in the same connection pattern as in Figs. 10.1 and 10.3, with the 4th pole switching gain resistors, R_{Gi}, in the differential preamp; a differential preamp comprised of operational amplifiers, U1, U2, and U3, using feedback resistors, R_F, for a total gain of $Gi = 1 + 2 * R_F/R_G$, as shown in Eq. (10.2); and a volume pot, P_V, feeding the output, Vo

$$Gi = \frac{Vo}{Vs} = \frac{R_{Gi} + 2 * R_F}{R_{Gi}} = 1 + \frac{2 * R_F}{R_{Gi}}, \quad R_{Gi} = \frac{2 * R_F}{Gi - 1}$$

$$Gi\# = \frac{Vs\,max}{Vsi}, \quad g = 1 + \frac{2 * R_F}{R_{Gimax}}, \quad Gi = g * Gi\# \qquad (10.2)$$

$$R_{Gi} = \frac{2 * R_F}{Gi - 1}$$

10.1.4   Embodiment 4: 4 Matched Pickups with a 4P6T Switch

See Fig. 10.8.

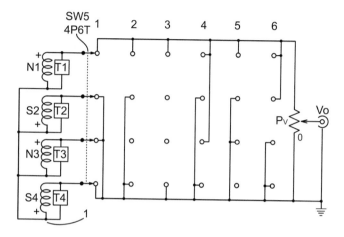

Fig. 10.8 From Fig. 9 of NPPA 16/139,027 (Baker, 2018b), showing 2 N-up pickups, N1 and N3, and 2 S-up pickups, S2 and S4, in positions 1–4 under the strings, connected to the common-connection point (1), connected in 6 throws of SW5 to produce signals corresponding to: $N1 + (S2 - N3 + S4)/3$, $N1 + (S2 + S4)/2$, $N1 + S4$, $(N1 + N3)/2 + (S2 + S4)/2$, $N1 + (S2 - N3)/2$, and $(N1 - S2)/2 + (S4 - N3)/2$. The top and bottom throws are the high and low output terminals, driving the single-ended output voltage, Vo, through volume pot, P_V. The standard output tone pot and capacitor have been replaced by circuits T1–T4, as in Fig. 10.4 or 10.7. Since there are 25 different humbucking tone circuits, using 2, 3, and 4 pickups each, and a total of 116 for the 16 different pole configurations, if the pickup magnets are reversible, this figure represents a very limited set of all the possible tone circuits

10.1.5 Embodiment 5: 4 Matched Pickups, or 2 Matched Humbuckers, with a 6P6T Switch and Differential Preamp

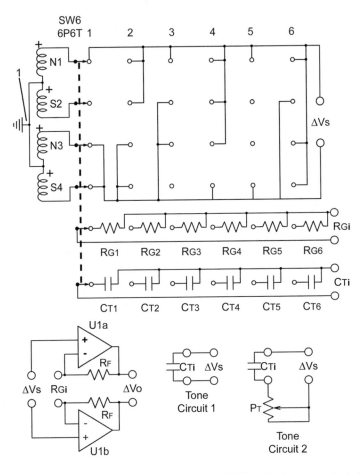

Fig. 10.9 From Fig. 13 of NPPA 16/139,027 (Baker, 2018b), showing 4 matched single-coil pickups, or 2 matched humbuckers, with coils N1, S2, N3, and S4, with 6P6T switch, using 4 poles to switch coils and a pole each to switch gain resistors, R_{Gi}, and tone capacitors, C_{Ti}. Note that the switch differential output, ΔVs, connects to three different circuits, the switch output, the differential amplifier, U1a and U1b, and one of two alternate tone circuits

The switch connections produce the strings signals relating to: $N1 + (S4 - N3)/2$, $(N1 - S2)/2 + (S4 - N3)/2$, $N1 + (S2 - N3)/2$, $(N1 - S2 + N3)/3 + S4$, $N1 + S4$ and $(N1 + N3 - S4)/3 + S2$, as shown in Table 10.3. If the inductance of a single coil is Lc, then the lumped inductance of 1 over 1 coils is 2Lc, 1 over 2 coils is 3Lc/2,

2 over 2 coils is Lc, and 1 over 3 coils (or vice versa) is 4Lc/3. Equation 10.3 shows how the roll-off frequency of the circuit is affected by the circuit inductance and the tone capacitance. If the roll-off frequency is to be the same for all 6 throws, then the product of the circuit lumped inductance times the tone capacitance must be a constant. In Eq. (10.3), Ca, Cb, Cc, and Cd are placeholders, and are adjusted to the value of Ca.

Table 10.3 From Table 12 in NPPA 16/139,027, results for Fig. 10.9, using corresponding mean frequency numbers from Table 9.3, showing the lumped circuit inductance for humbucking pairs, triples, and quads, for a matched pickup inductance of Lc

Throw	Pickup circuit signal	Mean freq (Hz)	Lumped circuit inductance
1	N1+ (S4 − N3)/2	1201.1	$3 * L_C/2$
2	(N1 − S2)/2 + (S4 − N3)/2	1006.8	L_C
3	N1+ (S2 − N3)/2	933.1	$3 * L_C/2$
4	(N1 − S2 + N3)/3 + S4	827.6	$4 * L_C/3$
5	N1 + S4	713.5	$2 * L_C$
6	(N1 + N3 − S4)/3 + S2	632.9	$4 * L_C/3$

$$2\pi f = \frac{1}{\sqrt{LC}} \Rightarrow Ca * LC = Cb * \frac{3 * LC}{2} = Cc * \frac{4 * LC}{3} = Cd * 2 * LC$$

$$Ca = C_{T2}, \quad Cb = \frac{2}{3}Ca = C_{T1} = C_{T3}, \qquad\qquad (10.3)$$

$$Cc = \frac{3}{4}Ca = C_{T4} = C_{T6}, \quad Cd = Ca/2 = C_{T5}$$

Nevertheless, this is not an efficient way to get all of the tonality available from 4 matched pickups, even if a plug board like that in Fig. 10.2 is used. For 4 pickups and up, electromechanical switches are too limited. We need another approach.

10.2   Digital Switching Systems for Common-Point Connection Circuits

The possible results of Tables 8.4 and 8.5, of so many more configurations and tones than electromechanical switches can control, justify the use of digitally controlled analog switches. Micro-power micro-controllers (uC) offer display, user interfaces, control and longer battery life, but few if any have the arithmetic processing units with the necessary trigonometric functions to calculate Fast Fourier transforms, which might be used to order tones. It may be necessary to add math processing units (MPUs). With such capability, and not yet fully determined algorithm for determining timbre and tone from strummed strings, it should be possible to offer the musician a user interface with a simple one-switch to one-swipe control to shift progressively from bright to warm tones and back without the musician ever needed to know which pickups are used in what configurations. In this disclosure, the mean frequency of six strummed strings is used as an example of the order of tone, which will likely be superseded by other measures. Nevertheless, the system architecture that will allow such measures and control will remain relatively constant for a while.

10.2.1 Embodiment 6: J N-up and K S-up Coils w/ Digital Control of SMD Analog Switches

Suppose that we have J number of N-up pickup coils and K number of S-up pickup coils, and we have chosen to use the common connection point switching system, where one terminal of each coil, regardless of magnetic pole direction (or electric pole for other sensors), are connected to a single point according to the same phase of external hum. In this switching system, there are 3 choices, or 3 states, for the other terminals of each coil to be connected by the switching circuit: (1) connected to the low output terminal of the switching system; (2) connected to the high output terminal of the switching system; or (3) not connected to either terminal. There is also the choice of how the ground is connected in the switching system, according to Rule 3. It is connected either to the low output terminal, or to the pickup common connection point. It is also possible to break the Rule, and ground both the common pickup connection and the low output terminal, so as to isolate the output of just one coil connected to the high output, for tuning and measurement purposes.

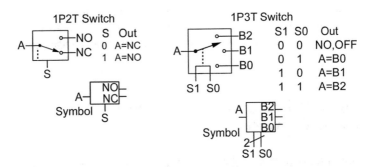

Fig. 10.10 From Fig. 14, NPPA 16/139,027. Two commercially available digitally controlled analog switches, 1P2T and 1P3T with on-resistance less than 2 Ω, supply voltages in the range of 1.65–4.3 V, and price <$1. The A terminal is the pole and the NO, NC, B0, B1, and B2 terminals are the throws. The S, S0, and S1 terminals are the digital controls

For this we need digitally controlled solid-state analog signal switches to reach the full potential of a switching system with more than 3 coils. And even for just 3 coils, it would be useful if the magnets are reversible. Figure 10.10 shows two such switches, a single-pole double-throw switch, and a single-pole triple-throw switch, with the additional state of no connection at all. The 1P2T switch in has a normally closed (NC) connection to the single pole, A, when the digital control, S, is at a low or binary "0" state. When $S = 1$, A is connected to the normally open connection, NO. The 1P2T switch can be used to connect the system analog ground to either the low output terminal or to connect the low output terminal to the pickup common connection, depending upon whether the amplifier at the switching system output is single-ended or differential. Or it can be used to switch tone capacitors.

The 1P3T switch has one pole, A, which is connected as shown in the table for the digital inputs, (S1, S0), to B0, B1, B2 or nothing, an open circuit, NO. When the digital state of (S1, S0) = (0, 0), the A terminal is connected to nothing, like an open circuit. When (S1, S0) = (0, 1), A is connected to B0, which can be the low output terminal. When it is (S1, S0) = (1, 0), A is connected to B1, which can be the high output terminal. When (S1, S0) = (1, 1), A is connected to B2, which can be the pickup common connection, thus shorting out the coil.

While it is possible to use a digitally controlled analog cross-point switch, they can come as large DIP chips, with more than a score of pins, or require supply voltages in excess of 5 V, or have contact resistances of tens of ohms. A cross-point switch typically addresses only one contact at a time, requiring addressing and data strobing for each separate connection. For a 6×8 cross-point switch (should one exist), used with four coils, a set of gain resistors and a set of tone capacitors, as in Fig. 10.9, there are $6 \times 8 = 48$ different cross-connections that have to be set individually by addressing.

The switches in Fig. 10.10 have only 1 or 2 bits of digital control, which can be the output lines of a micro-controller. In some cases, it may be advisable to add latches, if those uC lines are also used for other functions. The switches are small surface mount devices, often costing less than a dollar (US) each, with contact resistances down to about 1 Ω.

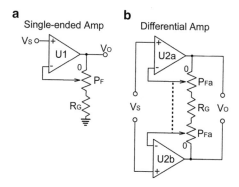

Fig. 10.11 From Fig. 15, NPPA 16/139,027. In (**a**) shows a digital pot, P_F, used to control gain as the feedback and part of the gain resistor, R_G. In (**b**) shows a dual digital pot, with independent sections P_{Fa} and P_{Fb}, used with gain resistor, to control the gain of a differential amplifier. The "0" symbols indicate the digital wiper position for the lowest digital input

With 4 coils, there are as many as 25 possible circuits requiring as many as 25 gain resistors to equalize the signal voltages. Or, alternatively and more efficiently, since a micro-controller is now available, digital pots can be used to set gain. Figure 10.11 shows a single-ended amplifier (a) and a differential amplifier (b), with feedback digital pots P_F, P_{Fa}, and P_{Fb}, op-amps U1, U2a, and U2b, and 2 gain resistors, R_G. The digital feedback pots can be 100k with 256 equal steps. Equations 10.4a and 10.4b show the equations for the pots and circuit gain. The dotted line between P_{Fa} and P_{Fb} indicates that they must be set to the same wiper positions in tandem to keep the output balanced about signal ground. Some digital pots come 2 to a chip. The output of the differential amp can be either differential, as shown, for use in further signal conditioning, or use the single-ended output structure of U3, R_F and P_V in Fig. 10.7.

$$\text{Resistance from ``0'' to } P_F \text{ wiper} = \frac{n * P_F}{256}$$

$$\text{Resistance from } P_F \text{ wiper to other pot end} = \frac{(256 - n) * P_F}{256} \tag{10.4a}$$

$$\text{Single-ended gain} = \frac{V_O}{V_S} = G = \frac{R_G + P_F}{R_G + P_F * \dfrac{256 - n}{256}}$$

$$\text{Differential gain} = \frac{V_O}{V_S} = G = \frac{R_G + 2 * P_F}{R_G + 2 * P_F * \dfrac{256 - n}{256}} \tag{10.4b}$$

Calculations elsewhere, using the resistance granularity of digital pots, indicate that using digital pots to set gain in Fig. 10.11a, with 256 equal resistance steps, $P_F = 100k$ and $R_G = 5.1k$, can equalize the relative amplitudes in Tables 9.6 and 10.2 within a range of $\pm 5\%$, over a gain range of $G = 1.0$–20.6. Digital pots typically have a serial interface comprised of 3 lines. For 4 coils, there are only 4 different lumped circuit inductances. So only 4 tone capacitors are needed to compensate for those differences, possible with 4 of the 1P2T switches, requiring 4 lines of digital control, or 1 of the 1P3T switches and 1 a1P2T, using 3 lines of digital control, or 2 of the 1P3T switches, using 4 lines of digital control.

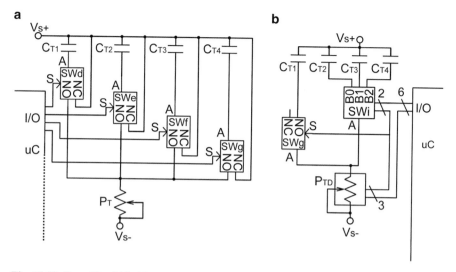

Fig. 10.12 From Fig. 16 in NPPA 16/139,027, showing two digitally controlled versions of tone capacitor selection for tone control at the input or output of Fig. 10.11a, b. (**a**) Shows a microcontroller using 4 I/O lines to control 4 1P2T analog switches to put any combination of 4 capacitors in series with a manual tone pot, P_T. (**b**) Shows a uC using 6 I/O lines to put combinations of 4 tone capacitors in series with a digitally controlled pot, P_{TD}

Figure 10.12 shows these alternatives, with 4 1P2T switches in Fig. 10.12a, and a single 1P3T switch with a single 1P2T switch in Fig. 10.12b, driven directly by the digital I/O lines of a micro-controller. In (a), the tone pot, P_T, is manual, and in (b), it is digital, P_{TD}, with 3 control lines going to the uC I/O. Either pot could be used in either side, depending on the overall circuit design. The circuit in (a) can produce 15 possible tone capacitances, or none, by connecting 0, 1, 2, 3, or 4 capacitors in parallel. The circuit in (b) can produce only 6, comprised of one of C_{T2}, C_{T3}, or C_{T4}, with or without C_{T1} in parallel. Table 10.4 shows the number of uC input/output lines needed for 4 coils, according to the circuits in Figs. 10.10–10.13. It may be advisable to use addressing and latches if some of these lines are to be used for other functions, such as User Controls and Displays.

Table 10.4 From Table 13 of NPPA 16/139,027, showing numbers of uC I/O lines needed for 4 coils in Fig. 10.13

			Min	Max
4 coils	4 1P3T	4 1P3T	8	8
4 tone caps	2 1P3T	4 1P2T	4	4
Tone pot	Manual	Digital	0	3
Single-ended or diff amp	1 Dig pot	2 Dig pots	3	6
Volume pot	Manual	Digital	0	3
Total			15	24

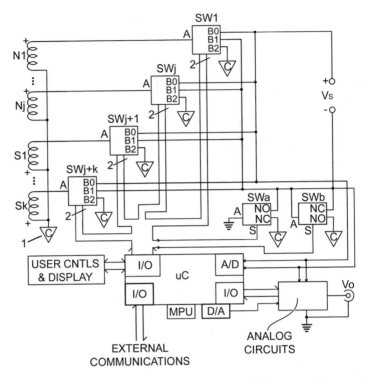

Fig. 10.13 From Fig. 17 in NPPA 16/139,017. Analog switching system for common-point connection pickups circuits, using a micro-controller. A C in a triangle designates the common point (1). *J* number of N-up matched single-coil pickups and *K* number of S-up pickups connect to common point with string signal polarities as shown by the "+" signs. Digitally controlled analog switches SW1 to SW$j + k$ are controlled the by micro-controller (uC) and switch the other terminals of the pickups to either Vs+ (B0), Vs− (B1), the common point (B2) or to no connection. Vs is the switching output. Switch SWa connects the ground to either the common point connection (default) for a differential output at Vs, or to Vs− for a single-ended output. Switch SWb shorts Vs− to the common point connection, so that just the pickups connected to Vs+ show at the output, either for maintenance/test functions, or for providing non-humbucking outputs of single pickups and other parallel combinations. The output, Vs, feeds into an analog-to-digital converter (A/D) in the uC for test functions and to take fast Fourier transforms of the output signal, which may be facilitated by a math processing unit (MPU) located either on or off the uC. The output also feeds into more traditional analog circuits, such as tone control, volume control, and distortion, which feeds into the instrument output, Vo. A digital-to-analog converter (D/A), either on or off the uC, and input-output controls (I/O) feed into the analog circuits to allow audio tones generated from the reverse Fourier transforms of frequency spectra to be fed to the instrument output. The uC can accommodate external communications, such as USB and Bluetooth, or external flash memory, for test and control functions and for storage of signal sampling and spectral information. It can also interface to user controls and displays, for test and control functions, such as switching tones from bright to warm with a simple tone-shift control

Figure 10.13 shows a micro-controller architecture for switching the combinations of J number of N-up coils, N1 to Nj, with their negative phase terminals connected to the common connection point (1), also denoted by a "C" in a triangle, and K number of S-up coils, S1 to Sk, to the switching system output, Vs, and then on to the analog signal conditioning circuits and the guitar output, Vo. The intermediate coils are not shown. The User Cntls and Display and MPU sections are explained below, and the Analog Circuits section is made up of circuits from Figs. 10.1, 10.3, 10.7, 10.9, 10.11 and/or 10.12.

The outputs of the coils are switched by the respective 1P3T digital-analog switches, SW1 to SWj, and SWj + 1 to SWj + k. The intermediate switches are not shown. The 1P3T switches, as in Fig. 10.10, have a four-state output, leaving the A terminal normally open, or connected to the B0 terminal, or the B1 terminal, or the B2 terminals, which are shown reversed vertically from Fig. 10.10, to simplify the circuit. All the B0 switch terminals go to the high switch output terminal, Vs+; all of the B1 switch terminals go to the low switch output terminal, Vs−; and all of the B2 switch terminals go to the pickup common terminal (triangle-C). So for each of the (S1, S0) states, (S1, S0) = (0, 0) disconnects the coil from any other part of the circuit; (S1, S0) = (0, 1) connects the coil to Vs+; (S1, S0) = (1, 0) connects the coil to Vs−; and (S1, S0) = (1, 1) connects the coil to the common terminal, shorting it out. Whether shorting the coil out has any effect on the tonal outputs remains to be determined.

The two 1P2T switches, SWa and SWb, perform other functions. For S = 0, SWa connects the ground to the pickup common connections, making the switching output, Vs, suitable for connection to a differential amplifier in the Analog Circuits section (Figs. 10.1, 10.3, 10.7, 10.9, 10.11 and 10.12, with the Fig. 10.3 P_V output for Fig. 10.11a and Fig. 10.7 output for Fig. 10.11b). For S = 1, SWa connects the ground to Vs−, making Vs suitable for connection to a single-ended amplifier in the Analog Circuits section. Since the Analog Circuits section is not likely to be switchable between single-ended or differential amplifiers, SWa could be replaced by a set of jumpers performing the same function.

For S = 0 (a separate control line from SWa), SWb shorts itself out and has no function, but for S = 1, it connects Vs− to the pickup common connection point (1), allowing the output of a single pickup coil, or a set of parallel pickup coils, connected to Vs+ to be fed to the Analog Circuits section. This will be useful for measuring or tuning single coils. The Analog Circuits section is taken to contain all the analog signal circuits. Figure 10.13 shows sensor and control lines between it and the micro-controller, uC, to handle such functions as the switching of tone capacitors and amplifier gains.

The micro-controller, uC, is shown with two-way digital connections to the User Controls and Display (adequately defined in NPPA 15/616,396); one-way control connections to 1P3T switches SW1 to SWj + k; one-way control connections to SWa and SWb; one-way connections from the switching system output, Vs, to an internal analog-to-digital converter (A/D); two-way sense and control connections with the Analog Circuits section, and a Math Processor Unit (MPU). The MPU section can be either internal to the uC, if available, or an add-on co-processor. Either

way must be capable of at least 32-bit floating point operations on complex variables, having sufficient trig and/or other math functions to accomplish Fast Fourier
Transforms (FFTs).

Using start-stop signals from the Analog Section or the User Controls and
Display, the FFT section performs complex FFTs on such inputs as the six strummed
strings, as described in Sect. 9.1. The FFT section takes A/D information from the
audio signal, Vs, to generate the complex FFTs needed for Eq. (9.1). The complex
FFTs generated should have a resolution of at least 1 Hz, and a frequency range of at
least 0–4 kHz, preferably to 10 kHz, and adjustable in bandwidth. It will be
necessary to switch the pickups during the A/D signal collection to obtain nearly
simultaneous sequential measurements either of all the coils separately, and/or all the
coils in humbucking pairs, corrected for time delays according to Eq. (10.5), to
produce effectively simultaneous complex FFT spectra for the calculations in
Eq. (9.1), where t_0 is the time delay.

$$x(t - t_0) \Leftrightarrow X(f) * e^{-j2\pi f t_0}, \quad e^{-j2\pi f t_0} = \cos(2\pi f t_0) - j\sin(2\pi f t_0) \quad (10.5)$$

A digital-to-analog converter (D/A), which can be either internal in the uC, or an
external circuit, feeds the audio from inverse-FFT transformations of measured
signal spectra into the Analog Circuits section to help the user recall pickup circuit
tones and to make better decisions on any user-defined tone switching sequences.
From this information, the switched coil combinations can be ordered by mean
output frequency from bright to warm or warm to bright, as a first approximation of
the order of tones. Or set by user preference. The tones in signal output from the
switching system can be equalized in volume or tone, according to Eqs. 10.1–10.4a
and 10.4b, in the Analog Circuits section by variable gains set by the uC. Then the
user can use the User Controls and Display to shift monotonically from tone to tone
without having to specify the particular switched coil combination that produces it.

10.2.2 Embodiment 7: Digital Switching Without a Micro-Controller

If for some reason a uC will not be used, the switching circuit in Fig. 10.13 can be
controlled by a simple up-down switch and an up-down digital ripple counter using
the same number of ripple outputs as the number of desired circuits to be switched.
The same 1P3T solid-state switches can be used. Gain resistors and tone capacitors
can be switched from the same ripple counter control signals. Another up-down
switch and ripple counter can be used for switching tone capacitors, if desired. Here
again, the plug board from Fig. 10.2 can be useful, especially if more than 3 pickups
or reversible-pole pickups are used. It can also be adapted to many more than just
6 switched selections.

The single bit of each ripple output can be connected to multiple switch control lines (S, S0 and S1 in Fig. 10.13), with each connection set isolated from every other by something as simple as diode or transistor isolators. Some digital signal inverters will likely be necessary. In diode isolators, two or more diodes can be connected with all the anodes, or all the cathodes connected to the control line output from the ripple counter, and the other terminal to each of the relevant switch controls. The direction of the diode polarities depends only on whether the switch control lines have either pull-up resistors to be pulled down by the ripple output, or pull-down resistors to be pulled up. Schottky with a low forward voltage drops will work best. It's an old technique dating back to the diode-transistor logic (DTL) of the 1960s that still works. It's so old and simple that a Figure is not necessary to illustrate it.

References

Baker, D. L. (2016b). Acoustic-electric stringed instrument with improved body, electric pickup placement, pickup switching and electronic circuit, US Patent 9,401,134, 26 July 2016. Retrieved from https://patents.google.com/patent/US9401134B2/

Baker, D. L. (2017a). Humbucking switching arrangements and methods for stringed instrument pickups. US Patent Application 15/616,396, filed June 7, 2017, published as US-2018-0357993-A1, December 13, 2018.

Baker, D. L. (2018a). Single-coil pickup with reversible magnet & pole sensor, US Patent Application 15/917,389, 9 Mar 2018. Retrieved from https://www.researchgate.net/publication/331193192_Title_of_Invention_Single-Coil_Pickup_with_Reversible_Magnet_Pole_Sensor

Baker, D. L. (2018b). Means and methods for switching odd and even numbers of matched pickups to produce all humbucking tones, US Patent Application 16/139,027, 22 Sep 2018, published as US-2019-0057678-A1, Feb 21, 2019, granted as U.S. Patent 10,380,986, 08/13/2019. Retrieved from https://www.researchgate.net/publication/335728060_NPPA-16-139027-odd-even-HB-pu-ckts-2018-06-22

Chapter 11
Humbucking Basis Vectors: Tones Without Switching

11.1 Some Prior Art

So far as this author knows, there is no prior art in this area other than his own, at least not in the field of guitar pickups. I started working out the math in early October 2017. On October 19, 2017, I filed the U.S. Provisional Patent Application 62/574,705, Using Humbucking Basis Vectors for Generating Humbucking Tones from Two or More Matched Guitar Pickups. On December 15, 2017, I filed PPA 62/599,452, Means and Methods of Controlling Musical Instrument Vibration Pickup Tone and Volume in STU-Space. On October 10, 2018, I filed the Non-Provisional Patent Application 16/156,509, Means and methods for obtaining humbucking tones with variable gains. As of this writing, in February–March 2019, the NPPA 16/156,509 is in process, awaiting examination.

11.2 The Circuits Created by the Math

This kind of math, linear vector and matrix math, comes from the author's undergraduate courses in the 1960s. Perhaps the easiest way to grasp it is to look the basic circuit it defines and predicts. This skips a lot of development the author had to do and makes it look obvious and easy. Besides, the circuits may actually have come first and suggested the math.

© Springer Nature Switzerland AG 2020
D. L. Baker, *Sensor Circuits and Switching for Stringed Instruments*,
https://doi.org/10.1007/978-3-030-23124-8_11

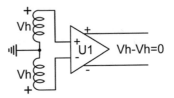

Fig. 11.1 Two matched pickup coils, either from a dual-coil humbucker, or two matched single-coil pickups, connected to a differential amplifier, U1, so that their equal hum voltages, Vh, cancel at the amplifier output. Derived from Fig. 1, Baker (2018c)

Figure 11.1 shows a basic humbucking circuit with two matched pickup coils, which should be familiar from the common connection point circuits of Chaps. 9 and 10. Humbucking does not depend upon the polarity of the magnetic field, which is not shown, but only upon the interconnections of the pickup coils so that the equal hum voltages in each of the coils cancels at the output. This could be the result if the magnets were replaced by an un-magnetized ferromagnetic material of relatively equal permittivity.

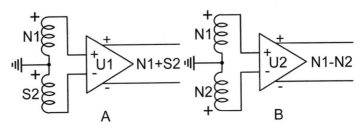

Fig. 11.2 Two matched pickup coils with the magnetic fields and string signal polarities shown. N means N-up, and S means S-up. From Fig. 1, Baker (2018c)

Figure 11.2a,b shows the same circuit, but with the pickup magnets labeled. We use here the arbitrary convention that the N-up coil string signal polarity coincides with the hum voltage, and the S-up signal polarity is opposite. So Fig. 11.2a produces the in-phase N1 + S2 output. And when the magnet is reversed in the pickup in position 2, Fig. 11.2b shows the contra-phase N1 − N2 output. Simple.

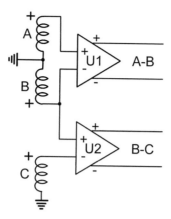

Fig. 11.3 Three matched pickup coils with the hum signals and polarities, A, B, and C, shown connected to fully differential amplifiers, U1 and U2, to produce zero hum outputs, A − B and B − C. From Fig. 2, Baker (2018c)

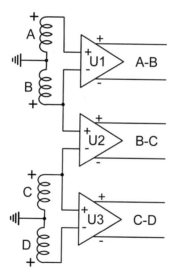

Fig. 11.4 Four matched pickup coils with the hum signals and polarities, A, B, C, and D, shown connected to fully differential amplifiers, U1, U2, and U3, to produce zero hum outputs, A − B, B − C, and C − D. From Fig. 3, Baker (2018c)

Figures 11.3 and 11.4 show how this type of circuit is scaled to 3 and 4 pickups, producing null outputs from the hum voltages, A to C or D. When the hum voltages are replaced by signal voltages, according to which magnetic pole is up toward the strings, the N-up pickups keep the same signal polarities and the S-up pickups reverse the signal polarities. So of A to D in Fig. 11.4 were converted to N1, S2, S3 and N4, the output signals for the differential amplifiers would be: (N1 + S2), (−S2 + S3), and (−S3 − N4).

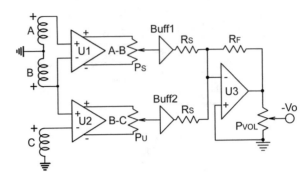

Fig. 11.5 Three matched pickup coils with the hum signals and polarities, A, B, C, with pots P_S and P_U acting as equation coefficients, s and u, added in the negative summer circuit, Buff1, Buff2, $2 \times R_S$, R_F and U3, with output volume pot P_{VOL}. From Fig. 6, Baker (2018c)

In Fig. 11.5, if U1 has unity gain, when the wiper of pot P_S is all the way up, the input of Buff1 sees the voltage $(A - B)/2$; when it is all the way down, Buff1 sees $-(A - B)/2 = (B - A)/2$. Likewise for P_U. Buff1 and Buff2 can be either unity gain buffers, or have another gain, as needed. The gain of the summer is $-R_F/R_S$. If the buffers are unity gain, and $R_F/R_S = 2$, then Vo $= -(s(A - B) + u(B - C)) = 0$. The gains of U1, U2, Buff1, Buff2, and R_F/R_S should be scaled so that no signal voltage clips on the circuit power supply levels. The summer can also be constructed so that Vo $= s(A - B) + u(B - C) = 0$.

When the string signals are considered, replacing A, B, and C with string signals according to N-up or S-up pickups, the convention of reversing the polarity of the S-up string signal must be followed, +Nx or an N-up pickup, and –Sx for an S-up pickup. Unless one is using the opposite convention, but let's not confuse things.

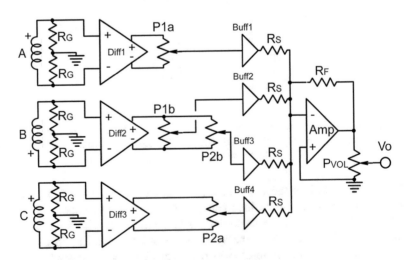

Fig. 11.6 An earlier version of Fig. 11.5, using differential amplifiers, Diff1 to Diff3, on each pickup, A, B, and C. Note that B is wired in opposite polarity, to generate hum signals proportional to $(A - B)$ and $(C - B)$. P1 and P2 are dual-gang pots with section a and b. The buffers, Buff1 to Buff4, resistors R_S and R_F, and the op-amp, Amp, for a negative summer, as in Fig. 11.5. From Fig. 3, Baker (2017c)

Figure 11.6 shows a different way to do the same thing with more components. But it does allow differential preamps to be mounted on individual pickups, which will help with common-mode noise reduction. To work as humbucking, either the B pickup has to be reversed in connection to the preamp, or the output connections from the preamp have to be reversed to the pots.

That makes the circuits in Figs. 11.1–11.5 more like "the elegant solution"; a term that MIT undergraduates used a lot back in the day.

11.3 And Now the Vector Math

These concepts are not necessarily presented in the order the author discovered and developed them, but in the order that makes the most tutorial sense. Recall the no-load pickup circuit equations from Chaps. 4, 7 and 8. In early October 2017, the author, working with humbucking triples, quads, and quints, generated no-load circuit equations. He began to see that supposedly different circuits could produce the same tone. On October 9th, his notes show that series and parallel circuits can be equivalent when the load resistance goes to infinity, or no-load open circuit, and that the number of distinct tones will be smaller than previously expected.

On the next page, he wrote:

"The advantage of this:

1. For $K = 4$, only 4 complex FFTs need be measured, 1 for each sensor/pickup.
2. The FFTs can be linearly combined according to the coefficient matrices to generate moments of freq. and order them from warm to bright.
3. Can use active sensors w/ a few simple gains and w/ phase reversal feeding into an analog cross-point switch.
4. If single-ended signals are used, the cross-point switch can be much simpler, w/K inputs and one output. Maybe not even a cross-pt sw, just a summer."

On October 10th, considering 5 matched coils in two humbuckers and a single-coil pickup, he wrote, "Is there a 5 × 5 matrix of vectors from which all humbucking circuits can be predicted w/ linear matrix operations?" The next day after brought Eq. (11.1).

$$[A \ B \ C \ D \ E] \begin{bmatrix} 1 & 1 & 1 & 1 \\ -1 & 0 & 0 & 0 \\ 0 & -1 & 0 & 0 \\ 0 & 0 & -1 & 0 \\ 0 & 0 & 0 & -1 \end{bmatrix} \begin{bmatrix} s \\ t \\ u \\ v \end{bmatrix}$$

$$= (s + t + u + v)A - sB - tC - uD - vE \qquad (11.1)$$

In this form, the scaled extension of Fig. 11.6 would have one gang from four control pots, Ps, Pt, Pu, and Pv, controlling the A signal, and the remaining pot gangs controlling the other four signals. In the original notes, A to E were string signals from all N-up pickups, producing humbucking contra-phase pairs. It works equally well for the hum signals to demonstrate humbucking. Eventually it dawned that the 1s in the top row of the 5×4 middle matrix need not all be in the top row, and for 3 pickups with the string signals, N1, S2, and S3, one can use either Eq. (11.2) or (11.3), with the conversion of coefficients in Eq. (11.4).

$$[\mathrm{N1} \quad -\mathrm{S2} \quad -\mathrm{S3}] \begin{bmatrix} 1 & 1 \\ -1 & 0 \\ 0 & -1 \end{bmatrix} \begin{bmatrix} s \\ t \end{bmatrix} = (s+t)\mathrm{N1} + s\,\mathrm{S2} + t\,\mathrm{S3} = \mathrm{Vo} \quad (11.2)$$

$$[\mathrm{N1} \quad -\mathrm{S2} \quad -\mathrm{S3}] \begin{bmatrix} 1 & 0 \\ -1 & 1 \\ 0 & -1 \end{bmatrix} \begin{bmatrix} \hat{s} \\ \hat{t} \end{bmatrix} = \hat{s}\,\mathrm{N1} + (\hat{s} - \hat{t})\,\mathrm{S2} + \hat{t}\,\mathrm{S3} = \mathrm{Vo} \quad (11.3)$$

$$\begin{bmatrix} \hat{s} \\ \hat{t} \end{bmatrix} = \begin{bmatrix} 1 & 1 \\ 0 & 1 \end{bmatrix} \begin{bmatrix} s \\ t \end{bmatrix} \Rightarrow \begin{matrix} \hat{s} = s + t \\ \hat{t} = t \end{matrix} \quad (11.4)$$

The astute reader will notice that these are not the usual square-matrix equations. They use $J \times (J - 1)$ matrices, because the need for two matched pickup coils to produce humbucking removes a degree of freedom. The columns of a $J \times J$ matrix in this application cannot be independent under humbucking rules. For J number of matched pickup coils, there can only be $J - 1$ number of humbucking coefficients. For the benefit of patent examiners, even though equations cannot be patented, that little detail makes this method non-obvious to prior art.

But, more importantly, the linear matrix equations that can apply to humbucking signals can also apply to their Fast Fourier Transforms. This makes it possible to take a digital sample series of individual humbucking pair signals, then their FFT transforms, multiply the FFTs by the humbucking coefficients, take the inverse FFT of the sum, and generate any signal in the n-space formed by the humbucking coefficients. More on this later.

As a matter of whimsy, the author called the coefficient space, STU-space, after the s, t, and u coefficients for a circuit with four matched pickup coils. Allegedly because it's much less intimidating than HARRY-space. Unfortunately, "t" is usually reserved for the time variable, so it had to become the much more stodgy SUV-space.

Recall what we learned of no-load tone equations in Chap. 4, derived from pickup circuit output equations. How do humbucking basis vector equations relate to that? Suppose that we have a hum signal equation in the form of a no-load output equation, for hum signals A_1, A_2, and A_3, with coefficients a_1, a_2, and a_3, as shown in Eq. (11.5).

$$Vo = a_1A_1 + a_2A_2 + a_3A_3 = 0$$
$$A_1 = A_2 = A_3 = V_{\text{HUM}} \Rightarrow a_1 + a_2 + a_3 = 0 \tag{11.5}$$

Equation (11.6) shows the equivalent humbucking basis vector equation, with coefficients s_1 and s_2.

$$Vo = s_1(A_1 - A_2) + s_2(A_2 - A_3) = s_1A_1 + (s_2 - s_1)A_2 - s_2A_3 = 0$$
$$A_1 = A_2 = A_3 = V_{\text{HUM}} \Rightarrow (s_1 + (s_2 - s_1) - s_2)V_{\text{HUM}} = 0 \tag{11.6}$$
$$s_1 + s_2 - s_1 - s_2 = 0, \quad \text{an apparently trivial solution}$$

Now consider the implications of the "trivial solution" in Eq. (11.6). It doesn't matter what the s_i coefficients are; putting the no-load output equation in terms of Eq. (11.6), and using the circuit configurations in Figs. 11.1–11.6, guarantees that the outputs are humbucking no matter what the values of the s_i coefficients may be. Not only can this approach produce all the tones of the all humbucking circuits developed all the way back through Chap. 2, it can produce all the continuous tones in between.

Equation (11.7) shows that there is no conflict between the (11.5) and (11.6). The last part of Eq. (11.5), $a_1 + a_2 + a_3 = 0$, requires a linear dependence of one coefficient on the other two, which removes a degree of freedom, just like requiring the matrices in Eqs. (11.1)–(11.3) to be $J \times (J - 1)$.

$$s_1 = a_1, \quad s_2 - s_1 = a_2, \quad -s_2 = a_3$$
$$a_1 + a_2 + a_3 = 0 \Rightarrow a_1 = -a_2 - a_3 \tag{11.7}$$
$$\therefore s_1 = -a_2 - a_3, \quad s_2 - s_1 = -a_3 - (-a_2 - a_3) = a_2$$

Here's another example. Recall from Sect. 7.2.2, according to Eq. (7.2), that when the a_i in Eq. (11.5) are all whole numbers, and Eq. (11.5) is an "output expression (or relation)" of string signals, having multiplied out a fractional denominator, the sum of a_i must be even for the expression to be humbucking. In Sect. 7.3.14, 18 of 24 five-pickup circuits in Fig. 7.5 were found to be humbucking, and 6 were not. Take the non-humbucking g(2 + 2 + 1) circuit with the output expression, Voc → A + B + 2C + 2D + 5E there is no way to set A to E to any combination of $\pm V_{\text{HUM}}$ to get a humbucking expression, with Voc = 0. Here, we use A to E instead of Ai to make it easier to see.

But as Table 7.7 shows, there are 8 ways to set up humbucking relations from the 18 circuits with even sums of coefficients. Take #1, Voc → A + B + C + D − 4E, where the equivalent coefficients $a_1 = a_2 = a_3 = a_4 = 1$, and $a_5 = -4$. That satisfies Eq. (11.5). Equation set (11.8) shows the relationships between the a_i, and s, u, v, and w for this example. Equation (11.9) shows the transform matrix to get from the a_i, i = 1–5, coefficients to the s, u, v, w coefficients. This form of transform matrix, with ones above the diagonal and zeros in and below the diagonal, works in general to get from the a_i coefficients to the s_i, or s–u–v..., coefficients. It can also be represented symbolically by $[H_{AS}]A = S$.

$$\text{Voc} \rightarrow A + B + C + D - 4E = s(A - B) + u(B - C) + v(C - D) + w(D - E)$$

$$\text{Voc} \rightarrow sA + (u - s)B + (v - u)C + (w - u)D - wE$$

$$s = 1, u - s = u - 1 = 1 \rightarrow u = 2, v - u = v - 2 = 1 \rightarrow v = 3,$$

$$w - v = w - 3 = 1 \rightarrow w = 4$$

$$\therefore \text{Voc} \rightarrow (A - B) + 2(B - C) + 3(C - D) + 4(D - E)$$

$$= A + B + C + D - 4E$$

$$(11.8)$$

$$\begin{bmatrix} 1 & 0 & 0 & 0 & 0 \\ 1 & 1 & 0 & 0 & 0 \\ 1 & 1 & 1 & 0 & 0 \\ 1 & 1 & 1 & 1 & 0 \end{bmatrix} \begin{bmatrix} 1 \\ 1 \\ 1 \\ 1 \\ -4 \end{bmatrix} = \begin{bmatrix} 1 \\ 2 \\ 3 \\ 4 \end{bmatrix} \quad \text{or} \quad [H_{AS}]A = S \qquad (11.9)$$

Notice that the coefficient of -4 in the column vector on the left side of Eq. (11.9) is the negative of the coefficient 4 in the right-side column vector. Recall Table 7.7, which listed 7 other humbucking relations for circuits of 5 matched single-coil pickups. In each case, the last coefficient of the left-side vector is the negative of the last coefficient of the right-side vector. The reader should try out the other 7 humbucking relations in Table 7.7.

There is also a reverse transformation matrix to Eq. (11.9), as shown below in Eq. (11.10). Equations (11.11) and (11.12) show that the order of multiplication of the transformation matrices H_{AS} and H_{SA} produce a standard identity matrix, I, in one case, but not the other.

$$\begin{bmatrix} 1 & 0 & 0 & 0 \\ -1 & 1 & 0 & 0 \\ 0 & -1 & 1 & 0 \\ 0 & 0 & -1 & 1 \\ 0 & 0 & 0 & -1 \end{bmatrix} \begin{bmatrix} 1 \\ 2 \\ 3 \\ 4 \end{bmatrix} = \begin{bmatrix} 1 \\ 1 \\ 1 \\ 1 \\ -4 \end{bmatrix} \quad \text{or} \quad [H_{SA}]S = A \qquad (11.10)$$

$$
\begin{bmatrix} 1 & 0 & 0 & 0 & 0 \\ 1 & 1 & 0 & 0 & 0 \\ 1 & 1 & 1 & 0 & 0 \\ 1 & 1 & 1 & 1 & 0 \end{bmatrix}
\begin{bmatrix} 1 & 0 & 0 & 0 \\ -1 & 1 & 0 & 0 \\ 0 & -1 & 1 & 0 \\ 0 & 0 & -1 & 1 \\ 0 & 0 & 0 & -1 \end{bmatrix}
=
\begin{bmatrix} 1 & 0 & 0 & 0 \\ 0 & 1 & 0 & 0 \\ 0 & 0 & 1 & 0 \\ 0 & 0 & 0 & 1 \end{bmatrix}
= I \text{ or } [H_{AS}][H_{SA}] = I
$$

$$(11.11)$$

$$
\begin{bmatrix} 1 & 0 & 0 & 0 \\ -1 & 1 & 0 & 0 \\ 0 & -1 & 1 & 0 \\ 0 & 0 & -1 & 1 \\ 0 & 0 & 0 & -1 \end{bmatrix}
\begin{bmatrix} 1 & 0 & 0 & 0 & 0 \\ 1 & 1 & 0 & 0 & 0 \\ 1 & 1 & 1 & 0 & 0 \\ 1 & 1 & 1 & 1 & 0 \end{bmatrix}
=
\begin{bmatrix} 1 & 0 & 0 & 0 & 0 \\ 0 & 1 & 0 & 0 & 0 \\ 0 & 0 & 1 & 0 & 0 \\ 0 & 0 & 0 & 1 & 0 \\ -1 & -1 & -1 & -1 & 0 \end{bmatrix} \neq I
$$

$$\text{or } [H_{SA}][H_{AS}] \neq I$$

$$(11.12)$$

But is it another kind of identity matrix? If we multiply it times the A matrix, we get back the A matrix, as shown in Eq. (11.13). We can call it the Humbucking Identity Matrix, I_{HB}, which works only upon A matrices for Voc humbucking relations.

$$
\begin{bmatrix} 1 & 0 & 0 & 0 & 0 \\ 0 & 1 & 0 & 0 & 0 \\ 0 & 0 & 1 & 0 & 0 \\ 0 & 0 & 0 & 1 & 0 \\ -1 & -1 & -1 & -1 & 0 \end{bmatrix}
\begin{bmatrix} 1 \\ 1 \\ 1 \\ 1 \\ -4 \end{bmatrix}
=
\begin{bmatrix} 1 \\ 1 \\ 1 \\ 1 \\ -4 \end{bmatrix}
$$

$$(11.13)$$

$$[H_{SA}][H_{AS}]A = [I_{HB}]A = A$$

Now, what happens when we rotate the coefficients in the A matrix? Let's take a different 5-pickup humbucking relation, #4 in Table 7.7, Voc $\rightarrow$ A − B − 2C − 3D +5E. Let's rotate the coefficients from $[1, -1, -2, -3, 5]$ to $[5, 1, -1, -3, -3]$ to $[-3, 5, 1, -1, -2]$ and so on, and apply the $[H_{AS}]$ transformation matrix to the resulting 5 × 5 $[A]$ matrix, as shown in Eq. (11.14). Rotating the coefficients in one direction has the same effect as rotating the pickup signals in the other direction. Exchanging coefficients in the Voc humbucking relation is the same as exchanging pickup positions in the circuit. Note that this does not result in rotating the entries of the resulting $[S]$ matrix. Doing so, and applying $[H_{SA}][S] = [A]$ will produce humbucking relations, but usually not those obtained from any series-parallel circuit topology. They will be "in-between" tones.

$$\begin{bmatrix} 1 & 0 & 0 & 0 & 0 \\ 1 & 1 & 0 & 0 & 0 \\ 1 & 1 & 1 & 0 & 0 \\ 1 & 1 & 1 & 1 & 0 \\ 1 & 1 & 1 & 1 & 0 \end{bmatrix} \begin{bmatrix} 1 & 5 & -3 & -2 & -1 \\ -1 & 1 & 5 & -3 & -2 \\ -2 & -1 & 1 & 5 & -3 \\ -3 & -2 & -1 & 1 & 5 \\ 5 & -3 & -2 & -1 & 1 \end{bmatrix} = \begin{bmatrix} 1 & 5 & -3 & -2 & -1 \\ 0 & 6 & 2 & -5 & -3 \\ -2 & 5 & 3 & 0 & -6 \\ -5 & 3 & 2 & 1 & -1 \end{bmatrix}$$

$$\text{or } [H_{AS}][A] = [S]$$

(11.14)

Table 11.1 shows the SUV-space coefficients for the discovered humbucking relations for series-parallel humbucking circuits for $J = 2, 3, 4, 5$, and 6 matched pickups, pairs, triples, quads, quints, and hexes, or for matched dual-coil pickups taken as two matched single coils. The starred hexes are provided from work not yet presented in this book. The author invites the reader to do and check the work. Note that for $J = 2$–6 matched pickups, there are 1, 1, 3, 8, and 20 different humbucking relations, respectively. In this field, the number of possibilities tends to go up exponentially with the number of matched pickups. The reader now has the tools to calculate how many combinations of pickups, by moving the individual pickups to other circuit positions can be constructed. But this approach makes that moot. All of those tones can be produced by changing SUV coefficients, as shown in Eq. (11.14), which can also produce all of the tones in between.

Note, however, that as more pickups are added, they are close together, and more circuits are required to change the SUV coefficients. There may be a point of diminishing returns, where the gradations in tone become so fine and indistinguishable that adding more pickups and circuits to a guitar becomes too much cost and trouble. But for larger instruments, like pianos, especially where no frets take up space on the strings, that point may be at a higher number of pickups and circuits (Table 11.1).

Table 11.1 Conversion of humbucking relation coefficients to SUV-space coefficients

Humbucking relation	SUV-space coefficients					Sum sq coef
	s	u	v	w	x	
Pairs						
A − B	1					1
Triples						
A + B − 2C	1	2				5
Quads						
A + B − C − D	1	2	1			6
A − B + 2C − 2D	1	0	2			5
A + B + C − 3D	1	2	3			14
Quints						
A + B + C + D − 4E	1	2	3	4		30
A + B − C + 3D − 4E	1	2	1	4		22
A + B − 2C + 3D − 3E	1	2	0	3		14
A − B + 2C + 3D − 5E	1	0	2	5		30
A + B + 2C − 2D − 2E	1	2	4	2		25
A + B + C − D − 2E	1	2	3	2		18
2A + 2B + 2C − 3D − 3E	2	4	6	3		65
2A + 2B + 3C − 3D − 4E	2	4	7	4		85
Hexes						
A + B + C − D − E − F	1	2	3	2	1	19
A + B + C + D − 2E − 2F	1	2	3	4	2	34
A − B + 2C + 2D − 2E − 2F	1	0	2	4	2	25
A − B + 2C − 2D + 5E − 5F	1	0	2	0	5	30
2A − 2B + 4C − 4D + 5E − 5F	2	0	4	0	5	45
A − B + C − D + 4E − 4F	1	0	1	0	4	18
A + B + C + 2D − 2E − 3F	1	2	3	5	3	48
A + B + C + 3D − 3E − 3F	1	2	3	6	3	59
A + B + C − D + 2E − 4F	1	2	3	2	4	34
A + B − C − D + 4E − 4F	1	2	1	0	4	22
A + B + C − 3D + 4E − 4F	1	2	3	0	4	30
A + B + C + D + E − 5F	1	2	3	4	5	55
2A + 2B + 2C + 3D − 3E − 6F	2	4	6	9	6	173
2A + 2B + 3C + 3D − 4E − 6F	2	4	7	10	6	205
2A − 2B + 4C − 5D − 5E+6F	2	0	4	−1	−6	57
A + B − C + 3D + 3E − 7F	1	2	1	4	7	71
A + B + 2C − 2D + 5E − 7F	1	2	4	2	7	74
A − B − 2C − 3D − 3E+8F	1	0	−2	−5	−8	94
A + B − 2C − 3D − 5E+8F	1	2	0	−3	−8	78
A + B − 2C + 3D + 5E − 8F	1	2	0	3	8	78

The square root of the sum of the squares of the SUV coefficients is the divisor that converts the SUV coefficient vector to a unit vector in SUV-space. The humbucking hexes have been completed elsewhere and provided here for completeness

The full nature of hex-coil tones are not yet settled; there may be errors. Make your own calculations and check them thoroughly

11.4 The Properties of SUV-Space

First, let's redraw Fig. 11.5 in a more symbolic and functional fashion, as Fig. 11.7, replacing standard pots Ps and Pu with two 360°, continuous rotation gangs on the same pot, where θ is the angle of rotation.

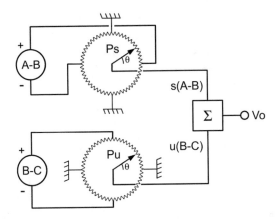

Fig. 11.7 A functional representation of Fig. 11.5, where the pickup signals $A - B$ and $B - C$ feed into continuous-rotation pot gangs, Ps and Pu, on the same pot, with angle of rotation, θ, and ground taps 90° off the signal voltage inputs. The resulting fractional inputs $s(A + B)$ and $u(B - C)$, where $-1 \leq s \leq 1$, and $-1 \leq u \leq 1$, feed into a summer to produce the output voltage, $Vo = s(A + B) + u(B - C)$

Now suppose that the gangs Ps and Pu have cosine and sin resistance functions, respectively, as shown in Fig. 11.8.

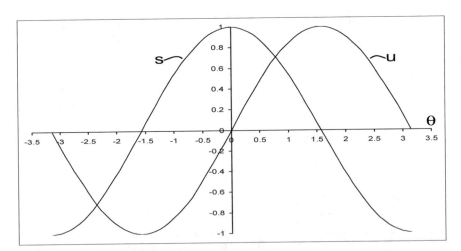

Fig. 11.8 Functions for Ps and Pu in Fig. 11.7 as cosine, $s = \cos(\theta)$, and sin, $u = \sin(\theta)$, pot gangs. From Fig. 3a in Baker (2017d) and Fig. 7 in Baker (2018c)

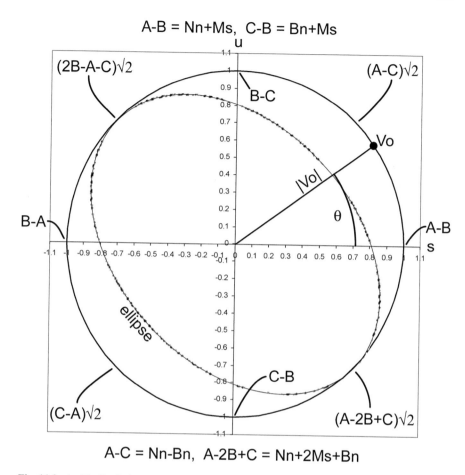

Fig. 11.9 An idealized plot of Vo. The circle shows the possible outputs of Vo for the pot rotation angle, θ, in Fig. 11.7, if the string signals A − B and B − C are equal in amplitude and have no phase interactions. If A is a N-up neck pickup, Nn, B is a S-up middle pickup, with signal −Ms, and C is a N-up bridge pickup, Bn, then the ellipse is much more likely, since the in-phase humbucking triple of Nn + 2Ms + Bn is likely to be louder than the contra-phase humbucking pair of Nn − Bn. From Fig. 3b in Baker (2017d) and derived from Fig. 8 in Baker (2018c)

As the captions of Figs. 11.8 and 11.9 relate, if pots Ps and Pu are continuous-rotation cosine and sin gangs on a single pot, with rotation angle θ, then one can expect the output, Vo, in Fig. 11.7, to behave something like Fig. 11.9. Anywhere along the line labeled IVol, from the origin to the signal, the tone does not change, only the amplitude. Therefore, 180° around the circle, the string signal is merely the opposite polarity. Note the signal composition and amplitudes around the circle, assuming that the string signals A − B and B − C have equal amplitude and no phase interactions (rather unlikely, but just for argument).

Suppose that the string signals A, B, and C represent pickups/signals, Nn, −Ms, and Bn, from an N-up neck pickup, an S-up middle pickup, and an N-up bridge pickup, and that all the pickups are matched single-coil pickups, just about the only kind allowed in this chapter. At $\theta = 0°$, the humbucking pair signal is A − B, or Nn + Ms., and marked around the rest of the circle are the humbucking contra-phase signal (Nn − Bn)√2, the negative phase humbucking pair signal −(Bn + Ms), the negative humbucking triple signal, −(Nn + 2Ms + Bn)√2, the negative signal, −(Nn + Ms), the negative contra-phase signal, −(Nn − Bn), the humbucking pair signal, Ms + Bn, and humbucking triple signal, (Nn + 2Ms + Bn)√2.

So only half the circle is needed to produce all the possible continuous-tone humbucking signals from 3 matched single-coil pickups. This will also hold for any of the other 3 possible pole configurations, including the all-contra-phase set where the pickups are either all N-up or all S-up.

This approach has limitations.

We know from experience that a bright contra-phase humbucking pair signal, like Nn-Bn, has less amplitude than a warm in-phase signal, like (Nn + 2Ms + Bn)/2 (Sects. 7.3.3 and 8.2.2). And in Fig. 11.9, the humbucking triple signal has a coefficient of 1/√2, which is larger than ½. Therefore, it may also have more amplitude that the humbucking pair signals, Nn + Ms and Ms + Bn. So the ellipse in Fig. 11.9 is closer to the truth, which would have to be measured in any case and may be even lumpier than the ellipse due to phase interactions between the pickup signals.

Note that only half of the circle or ellipse is needed to cover all the possible tones, but it takes only ¼ of the circle or ellipse to get from the presumably warmest tone, the humbucking triple, proportional to Nn + 2Ms + Bn, to the likely brightest tone, Nn − Bn. In one direction from the brightest tone, it and the humbucking triple combine with Nn + Ms, in the other, they combine with −(Ms + Bn). And the likely differences in phase and amplitude bias those shifts. So it is not possible with a single dual-gang rotary control to go monotonically from the brightest to the warmest tones. And especially not possible to do so with all the tones having the same amplitude. It will be a complex mix.

11.5 Analog Methods: Extending the Sin-Cosine Pot Approach to *J* > 3 Matched Pickups

However, it is equally obvious that for $J = 3$ matched single-coil pickups, all of the possible tones can be covered with just one control. We will see that for J number of pickups, it is possible to cover all the possible the tones with $J − 2$ number of controls. Abstracting Figs. 11.4 and 11.5 a bit farther, Figs. 11.10 and 11.1 show alternative sin–cosine controls for $J = 5$ pickups (or 2 dual-coil humbuckers and a single-coil pickup matched to the humbucker coils), using the trigonometric identities in Eq. (11.15). Although the squares of sines and cosines do not appear in the figures, they come in when calculating the length $|Vo| =$ the square root of the sum of the squares of the SUV 4-space coordinates (s, u, v, w) necessary to define the tonal space of five pickups, according to Eq. (11.16)

$$\cos^2(\theta) + \sin^2(\theta) = 1$$
$$[\cos^2(\theta) + \sin^2(\theta)]\cos^2(\lambda) + [\cos^2(\phi) + \sin^2(\phi)]\sin^2(\lambda) = 1 \qquad (11.15)$$
$$[[\cos^2(\theta) + \sin^2(\theta)]\cos^2(\phi) + \sin^2(\phi)]\cos^2(\lambda) + \sin^2(\lambda) = 1$$

$$Vo = s(A - B) + u(B - C) + v(C - D) + w(D - E)$$
$$|Vo| = \sqrt{s^2 + u^2 + v^2 + w^2} \qquad (11.16)$$

assuming the $A - B$, $B - C$, $C - D$, and $D - E$ are equal in magnitude to 1, for the purpose of argument.

$$Vo = s(A - B) + u(B - C) + v(C - D) + w(D - E)$$
$$Vo = \cos(\theta)\cos(\lambda)(A - B) + \sin(\theta)\cos(\lambda)(B - C) \qquad (11.17)$$
$$+ \cos(\phi)\sin(\lambda)(C - D) + \sin(\phi)\sin(\lambda)(D - E)$$

$$Vo = s(A - B) + u(B - C) + v(C - D) + w(D - E)$$
$$Vo = \cos(\theta)\cos(\phi)\cos(\lambda)(A - B) + \sin(\theta)\cos(\phi)\cos(\lambda)(B - C) \qquad (11.18)$$
$$+ \sin(\phi)\cos(\lambda)(C - D) + \sin(\lambda)(D - E)$$

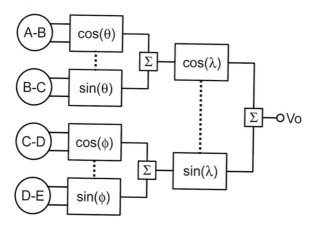

Fig. 11.10 A symbolic block diagram of the sine/cosine controls for 5 matched single-coil pickups with placeholders, A, B, C, D, and E, for the pickups, hum signals and string signals, using the second trig identity in Eq. (11.15). The angles, θ, ϕ, and λ, and the dotted lines indicate 3 dual-ganged sine/cosine controls providing the coefficients for the output equation, Eq. (11.17). The differential signal sources are $A - B$, $B - C$, $C - D$, and $D - E$, feeding into 2 sine/cosine ganged controls, which are summed and fed into a third sine/cosine ganged control, and summed again for the output, Vo

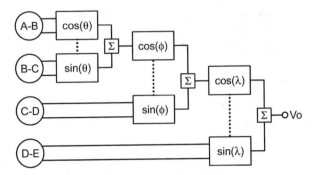

Fig. 11.11 An alternative method of summing humbucking signals, A − B, B − C, C − D, and D − E, using 3 ganged sine/cosine controls and 3 summers, with the output equation in Eq. (11.18)

The methods in Figs. 11.10 and 11.11 should make scaling to more or less pickups at least visually obvious. This author does not have as good or concise way of describing this scaling verbally. Note that both methods in Figs. 11.10 and 11.11 are the same for 3 pickups. Note that while a 3-coil guitar needs just one control to move on a circle or ellipse through $s–u$-space; 4 pickups will need 2 controls to move on a sphere or ellipsoid through $s–u–v$-space; and as shown, 5 pickups need 3 controls to move on a hyper-sphere or hyper-ellipsoid through $s–u–v–w$-space, and so on. And remember that when actually measured, they might be lumpy ellipsoids, more like potatoids.

Whether one or the other of the methods in Figs. 11.10 and 11.11 is better depends upon the trade-offs in the actual design and performance of the circuits. Equation (11.18) will likely have more error from multiplied sine/cosine terms than Eq. (11.17). But it may be easier to implement with components because of Fig. 11.12, which shows a way to use a standard one-turn, non-continuous rotation, sine/cosine-ganged pot.

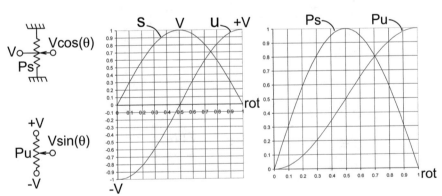

Fig. 11.12 Single-turn dual ganged sine/cosine pot and curves. The cosine gang has a center tap to apply $+V$, with the ends grounded, and the sine pot has no center tap, with $+V$ and $−V$ applied to the ends. The first graph shows the voltage outputs versus fraction rotation. The second graph shows the relative pot resistance at the pot wiper versus fractional rotation. All are associated with s and u coefficients. From Fig. 5 in Baker (2017d) and Fig. 9 in Baker (2018c)

The pots in Fig. 11.12 limit the SUV-space to the positive half of the s-axis, which is a half-circle for 3 coils, or a half-sphere for 4 coils, and so on. In Fig. 11.9, this goes from C − B to A − B to B − C with fractional rotation. So the presumably warmest tone, the humbucking triple, would sit at about ¼ rotation, and the presumably brightest tone, Nn − Bn, would sit at about ¾ rotation.

Figures 11.13 and 11.14 show the circuit embodiments of the dual-gang sine/cosine pot in Fig. 11.12 for 3 and 4 pickups, respectively, using the method of Fig. 11.11.

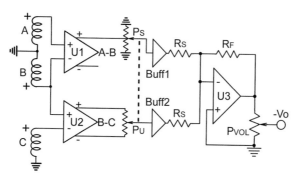

Fig. 11.13 Embodiment of a humbucking circuit for three matched pickups using the pots in Fig. 11.12. The "+" signs indicate hum voltage polarity. U1 and U2 are fully differential amplifiers. Ps and Pu are the gangs of a sine/cosine pot as in Fig. 11.12. Buff1, Buff2, (2)R_S, R_F and U3 comprise a signal summer. From Fig. 10 in Baker (2018c)

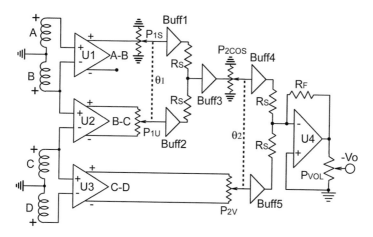

Fig. 11.14 The scaling of the circuit in Fig. 11.13 to 4 pickups, using the method of Fig. 11.11 and the pots of Fig. 11.12. The gangs P_{1S} and P_{1U} comprise one dual-gang sine/cosine pot with rotation angle θ_1. The gangs P_{2COS} and P_{2V} comprise a second dual-gang sine/cosine pot with rotation angle θ_2. Note that the sine sections, P_{1U} and P_{2V}, require fully differential inputs, while the cosine sections do not. As Eq. (11.18) indicates, the subscripts s, u, and v on the pots are not entirely accurate. From Fig. 11 in Baker (2018c)

11.6 Analog Methods: Replacing a Sine/Cosine-Ganged Pot with a 3-Gang Linear Pot

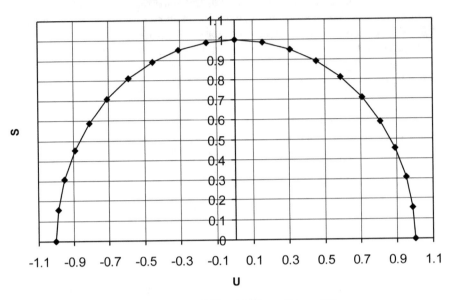

Fig. 11.15 Plot of $s = \cos(\theta)$ versus $u = \sin(\theta)$, with 20 equal steps of θ

For reference, Fig. 11.15 shows the plot of $s = \cos(\theta)$ versus $u = \sin(\theta)$ on a half circle. Note the evenly spaced steps in s–u-space. But sine/cosine pots are neither easy nor cheap to make. If one is willing to give up a little accuracy in trig identities in Eq. (11.15) and the vector length equation for |Vo| in Eq. (11.16), it is possible to use a 3-gang linear pot, as shown in Fig. 11.16.

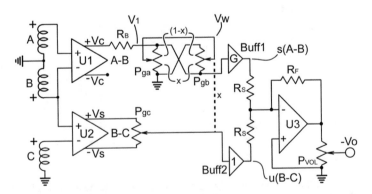

Fig. 11.16 The circuit of Fig. 11.13 with the sine/cosine dual-gang pot replaced by a 3-gang linear pot, Pg, with gangs Pga, Pgb, and Pgc. The "x" and "$1 - x$" refer to the portions of the total resistance of each gang below and above the gang wipers. Note that at the ends of wiper rotation, the wipers are grounded, and reach a maximum resistance at the middle of the rotation. From Figs. 17 and 19 in Baker (2017d) and Fig. 12 in Baker (2018c)

But sine and cosine are not the only functions for which $(s(x)^2 + u(x)^2) = 1$, where $0 \le x \le 1$ is the decimal fractional rotation of a single-turn pot with multiple gangs, having tapers $s(x)$ and $u(x)$. These functions can be simulated with a 3-gang linear pot. Figure 11.16 shows this circuit applied to Fig. 11.13. The linear pot gang, Pgc, of pot Pg in Fig. 11.16 replaces the sine-taper pot in Fig. 11.13, Pu, and simulates the scalar u. The circuit comprised of the resistor, R_B, and the two linear gangs, Pga and Pgc, of pot Pg, of resistance value, Rg, replaces the cosine-taper pot, Ps. The differential amplifiers, U1 and U2 are assumed to have a gain of 2. The plus output of U1, Vs, comes out of the 2-gang pot circuit on the wiper terminal as Vw. The combination of the resistor, R_B, the 2-gang circuit and the Buff1 with gain, G, simulates the scalar, s, as shown in Eq. (11.19).

$$\frac{V_1}{Vc} = \frac{2x(1-x)Rg}{2x(1-x)Rg + R_B}, \quad Vw = \frac{V_1}{2}$$

$$\left.\frac{V_1}{Vc}\right|_{x=1/2} = \frac{Rg}{Rg + 2R_B}, \quad \text{for} \quad G\left.\frac{V_1}{Vc}\right|_{x=1/2} = 1 \rightarrow G = \frac{Rg + 2R_B}{Rg} \tag{11.19}$$

$$G\frac{V_1}{Vc} = s(x) = \frac{2x(1-x)(Rg + 2R_B)}{2x(1-x)Rg + R_B}$$

Equation (11.19) shows the solutions to the circuit equations for R_B, Pga, Pgb, Vs, V_1, and Vw. In order for the simulation of the scalar, s, to have a range from 0 to 1, the gain, G, of Buff1 must be as shown. As noted in Fig. 12, the output of Buff1 simulates $s(A - B)$ and the output of Buff2 simulates $u(B - C)$. If $V = A - B = B - C = 1$, then Figs. 11.17 and 11.18 show the plots of $s(x)$, $u(x)$ and $(s^2(x) + u^2(x))^{1/2}$ as the tapers of the pseudo-cosine circuit and the linear Pgc with pot fractional rotation, x. Given the resistance value the gangs of the pot Pg, Rg, the value of R_B is changed by optimization until ε is minimized in Eq. (11.20). Then G is set in Eq. (11.19). For example, when Rg $= 10$k and $R_B = 2.923$k, ε optimizes to ± 0.0227, or less than 3% of scale.

$$1 - \sqrt{s^2(x) + u^2(x)} \le \pm\varepsilon \tag{11.20}$$

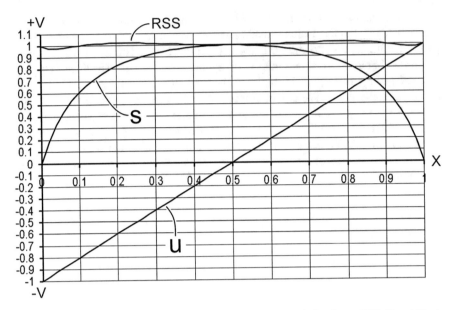

Fig. 11.17 Plot of normalized curves for s and u in Fig. 11.16, for Rs $= 2923$ Ω and Rpot-gang $= 10$ kΩ. RSS is the square root of the sum, $s^2 + u^2$. From Fig. 18 in Baker (2017d) and Fig. 13 in Baker (2018c)

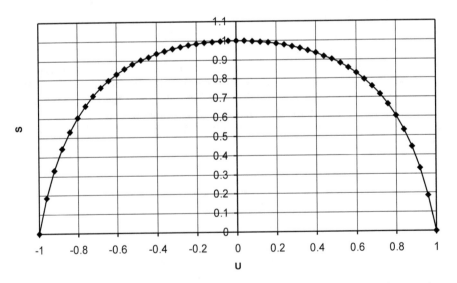

Fig. 11.18 Shows a half-circle plot of 51 points from Fig. 11.17 of s plotted against u, from $x = 0$ to 1, in 0.02 steps. From Fig. 18 in Baker (2017d) and Fig. 14 in Baker (2018c)

Note in Fig. 11.18 that the center of the range, around $x = 0.5$, has more resolution than for $x = 0$ or $x = 1$, due to matching a linear with a nonlinear curve. And how the points bunch together near $u = 0$, and form a flattened circle rather than a true circle. The flattening itself is sufficient to increase the resolution. Related to Fig. 11.9, this means that there is more resolution about the A − C or Nn + Ms tone than elsewhere. It is one of the trade-offs for using cheaper pots.

11.7 Digital Methods: Approximating Sine/Cosine Pots with Linear Digital Pots

With analog control, as we have seen, the tones can change continuously, but not in any order we might choose. But with a micro-processor or micro-controller, an analog-to-digital-converter, and digital pots and switching, the order of the tones can be calculated and set to anything the hardware and a software program can manage. First we look at using linear digital pots and changing the settings according to trig functions in software.

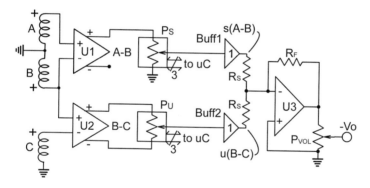

Fig. 11.19 Figures 11.13 and 11.16 with the sine and cosine gangs replaced by individual or dual linear digital pots, with 3 control lines going to a micro-controller. From Fig. 15 in Baker (2018c)

Figure 11.19 shows Figs. 11.13 and 11.16 with the analog pots replaced by digital pots, P_S and P_U, with 3-line digital serial control lines going to a micro-controller (uC), not shown. The fully differential amplifiers, U1 and U2, each have a gain of 2 and the buffers, Buff1 and Buff2 each have a gain of 1, providing and simulating signals $s(A − B)$ and $u(B − C)$ in concert with P_S and P_U. The micro-controller calculates the appropriate cosine (for Ps) and sine (for Pu) functions, and uploads them into the digital pots via the serial control lines. Depending on make and model, digital pots typically come with 32, 100, 128, or 256 resistance taps, linearly spaced to provide a total resistance across the pot of typically 5k, 10k, 50k, or 100k-Ω.

For this example, we will assume digital pot with 256 resistance taps. In this case, x as a decimal fractional rotation number from 0 to 1 has no meaning. The numbers 0 and 255 correspond to the ends of the pot, zero resistance to full resistance on the wiper. The internal resistor is divided into 255 nominally equal elements, and an 8-bit binary number, from 00000000 to 11111111 binary, or from 0 to 255 decimal, determines which tap is set. The pot either has a register which holds the number, or an up-down counter which moves the wiper up and down one position. The convention used here makes $s = \cos(\theta)$ and $u = \sin(\theta)$ for $-\pi/2 \leq \theta \leq \pi/2$, with $0 \leq s \leq 1$ and $-1 \leq u \leq 1$. So s maps onto $0 \leq Ns \leq 255$, and u maps onto $0 \leq Nu \leq 255$.

But while we want $s = \cos(\theta)$ and $u = \sin(\theta)$, where $-\pi/2 \leq \theta \leq \pi/2$, the digital pot doesn't quite work that way. It has no angle θ, which can be mapped onto any physical rotation. The number that is fed to the pot to set it must be an integer from 0 to 255. But we can start from $0 \leq x \leq 1$, where x is "fractional rotation," convert to $0 \leq \theta \leq \pi$, shift the sine and cosine functions by $\pi/2$, and use Eq. (11.21) to get mapping functions for Ns and Nu. Int() is a common computer function that rounds to the next lowest integer. We add 0.5 to the variable to get it to round to the nearest integer. The results are shown in Figs. 11.20 and 11.21.

$$\text{Int}(y) = \text{integer} \leq y, \quad \text{a computer function}$$
$$\theta = \pi x, \quad 0 \leq x \leq 1, \quad x \text{ is "fractional rotation"}$$
$$Ns = \text{Int}(255 \sin(\theta) + 0.5) \tag{11.21}$$
$$Nu = \text{Int}(127.5 * (1 - \cos(\theta)) + 0.5)$$

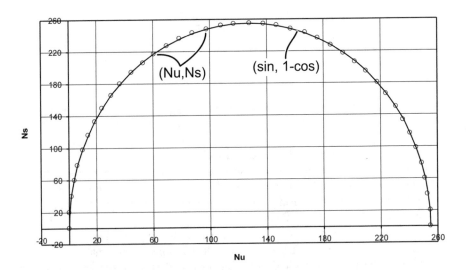

Fig. 11.20 Plots of Ns and Nu from Eq. (11.21). The circles are the positions of (Nu, Ns). The line is the 40 equal segments of $(255 * \sin(\pi x), 127.5 * (1 - \cos(\pi x)))$, where $0 \leq x \leq 1$ in 40 equal steps. The error in the radius of 127.5 is $-0.4164 < err < 0.392$, with root-mean-squared error of 0.242. That comes to a maximum absolute error in the radius of the curve of about 3.1%

$$Pu = Nu/255, \quad Ps = Ns/255, \quad Vs = Ps, \quad Vu = 2Pu - 1$$
$$Length = \sqrt{Vs^2 + Vu^2} \tag{11.22}$$

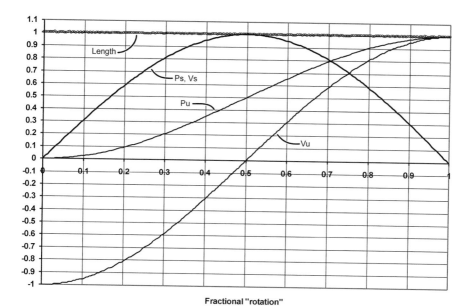

Fig. 11.21 Normalized plots of the digital potentiometer curves, Ps and Pu, obtained from Ns and Nu via Eq. (11.22), using steps of the "fractional rotation" x, $\Delta x = 1/255$. The resulting output curves, Vs and Vu, would result from Fig. 11.19, if $A - B = B - C = 1$, and the Length is the radius using Vs and Vu as cosine and sine in the first line of Eq. (11.15). Length is plotted as small circles to emphasize the error about the desired radius of 1. The error of (Length $- 1$) is less than ± 0.004. Compared to Fig. 11.20, the error in the radius of a circle described by the relative sine and cosine approximations seems to decrease with step size. From Fig. 10 in Baker (2017d) and Fig. 9 in Baker (2018c)

11.8 Digital Methods: Approximating Sine/Cosine with 5 Basic Functions

While ARM-based micro-controllers (uC) may have a trig functions, one very low power uC, which runs at about 100 uA (micro-amps) per MHz of clock rate, has 32-bit floating point arithmetic functions, add, subtract, multiply, divide and square root, but no trig functions or constant of Pi. If such a basic uC is chosen to run linear digital pots as sine/cosine pots, much less calculate Fast Fourier Transforms (FFTs), a good approximation is needed for trig functions. This author does not know exactly how math coprocessors work, only that $f(x)$ series expansions like Taylor and Maclaurin, are perfectly accurate at just one point, and increase in error with distance, x, from that point, requiring ever more terms in the expansion to reduce the error. This increases calculation time. So, if such a limited calculator must be used, finding methods that can use just 5 functions to approximate sine and cosine is advisable.

This requires two different orthogonal functions which can satisfy Eq. (11.20), but not necessary those in Embodiment 3. Equation (11.23) shows a set of functions, $s(x)$ and $u(x)$, which meet Eq. (11.20) with no error, and are orthogonal to each other (Figs. 11.22, 11.23, and 11.24).

$$\text{Int}(y) = \text{integer} \leq y$$

$$s = 1 - 4\left(x - \frac{1}{2}\right)^2, \quad u = \begin{cases} -\sqrt{1 - s^2}, & 0 \leq x < \frac{1}{2} \\ +\sqrt{1 - s^2}, & \frac{1}{2} \leq x \leq 1 \end{cases} \quad (11.23)$$

$$\text{Ns} = \text{Int}(255s + 0.5)$$

$$\text{Nu} = \text{Int}(127.5 * (1 + u) + 0.5)$$

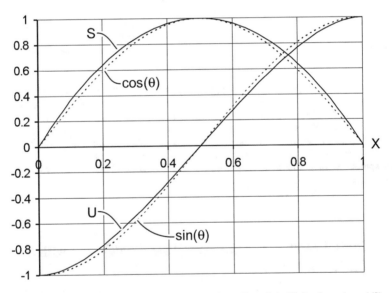

Fig. 11.22 Shows $s(x)$ and $u(x)$ in the solid lines, and $\cos(\theta)$ and $\sin(\theta)$ for $\theta = \pi(x - 1/2)$ as the dotted lines. The differences between s and $\cos(\theta)$ runs from 0 to 0.056, and the differences between u and $\sin(\theta)$ run from about -0.046 to $+0.046$. From Fig. 16 in Baker (2018c)

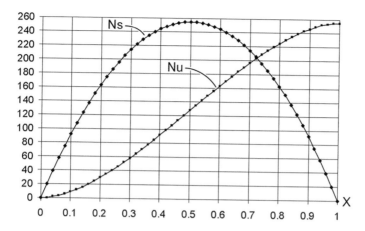

Fig. 11.23 Shows Ns and Nu from Eq. (11.23) for 51 values of *x* in steps of 0.02 from 0 to 1. These are a kind of pot-taper plot. Note that for $x = 0.5$, Ns = 255 and Nu = 128. The errors should be on the order of 1/255, plus the digital pot manufacturing errors. From Fig. 17 in Baker (2018c)

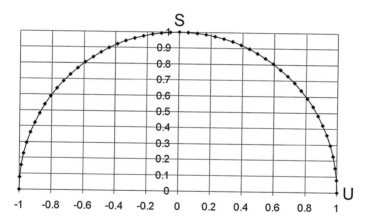

Fig. 11.24 Shows the *s*-versus-*u* half-circle plot for the same 51 values of *x*. Note that the distribution of points on the circle does not bunch like those for the pseudo-cosine-sine analog plot curves in Fig. 11.18. This is a much closer approximation to sine-cosine curves and is actually cheaper in digital pot part costs than analog potentiometers, not counting the circuit and uC costs. From Fig. 18 in Baker (2018c)

Equation (11.24) shows an even better function, plotted in Fig. 11.25, for $x = 0$ to 1 in steps of 0.01. The error for $s(x) - \cos(\theta(x))$ runs from 0 to -0.067 and for $u(x) - \sin(\theta(x))$ from -0.004 to $+0.004$.

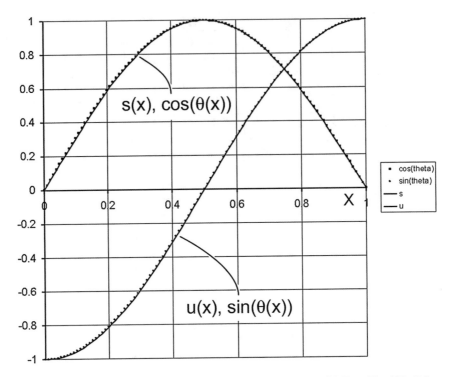

Fig. 11.25 Show $s(x)$, cosine, $u(x)$ and sine plotted according to Eq. (11.24) From Fig. 19 in Baker (2018c)

$$0 \le x < 1, \quad \theta(x) = \pi(x - 0.5)$$

$$s = 1 - 5\left(x - \frac{1}{2}\right)^2 + 4\left(x - \frac{1}{2}\right)^4, \quad u = \begin{cases} -\sqrt{1 - s^2}, & 0 \le x < \frac{1}{2} \\ +\sqrt{1 - s^2}, & \frac{1}{2} \le x \le 1 \end{cases} \quad (11.24)$$

The functions in Eqs. (11.23) and (11.24) suggest the candidates in Eqs. (11.25) and (11.26) to be substituted for sine and cosine in an FFT algorithm, when the uC has a floating point square root function, but no Pi constant or trig functions. In these cases, the variable of rotation is not $0 \le \theta < 2\pi$, but $0 \le x < 1$; the frequency argument of cosine changes from $(2\pi ft)$ to simply (ft), and the FFT algorithm must be adjusted to scale accordingly. Figure 11.26 shows the plots for $x = 0$ to 1.5, step 0.01. The error in Sxm-sin is -0.00672 to 0.00672 and the error in Cxm-cos is -0.004 to 0.004. Note how the scaling has changed between Eqs. (11.24) and (11.25).

$$xm = x \text{ modulo } 1, \quad \theta(x) = 2\pi x$$

$$\sin(\theta(x)) \approx S_{xm} = \begin{cases} 0 \le xm \le 0.5, & 1 - 5(2*xm - 0.5)^2 + 4(2*xm - 0.5)^4 \\ 0.5 < xm < 1, & -\left(1 - 5(2*xm - 1.5)^2 + 4(2*xm - 1.5)^4\right) \end{cases}$$

$$\cos(\theta(x)) \approx C_{xm} = \begin{cases} 0.25 < xm < 0.75, & -\sqrt{1 - S_{xm}^2} \\ \text{else}, & \sqrt{1 - S_{xm}^2} \end{cases}$$

$$(11.25)$$

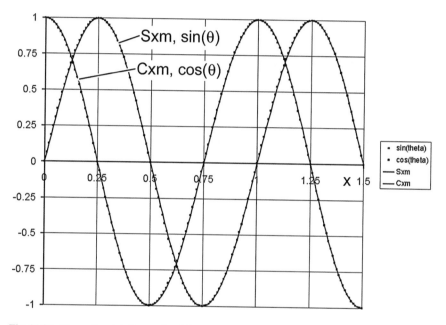

Fig. 11.26 Shows the plots for $x = 0$ to 1.5, step 0.01. The error in Sxm-sin is -0.00672 to 0.00672 and the error in Cxm-cos is -0.004 to 0.004. Note how the scaling has changed between Eqs. (11.24) and (11.25) from $(x - 0.5)$ to $(2x - 0.5)$, which is necessary to fit a full cycle into $0 \le x < 1$. From Fig. 20 in Baker (2018c)

$$xm = x \text{ modulo } 1, \quad xm2 = xm \text{ modulo } 0.5$$

$$a = (xm2 - 0.25)^2$$

$$S_{xm\text{-corr}} = \begin{cases} 0 \le xm \le 0.5, & 0.2629467*a + 0.7071068*a^2 - 78.62807*a^3 \\ 0.5 < xm < 1, & -(0.2629467*a + 0.7071068*a^2 - 78.62807*a^3) \end{cases}$$

$$S_{xm} = S_{xm} + S_{xm\text{-corr}}$$

$$(11.26)$$

Equation (11.26) shows an added correction to Sxm, prior to calculating Cxm, which reduces the error to less than $\pm 1.5e{-}6$ for Sxm, and less than $\pm 1.4e{-}5$ for Cxm. The precision of the coefficients is consistent with IEEE 754 32-bit floating

point arithmetic. Listing 11.1 shows a Fortran-like subroutine to calculate the sine-
and cosine-approximation return variables SXM and CXM from X and NORD. For
NORD = 0, a re-scaled Eq. (11.23) is calculated, for NORD = 1, Eq. (11.25) is
calculated, and for NORD = 2, the correction in Eq. (11.26) is added before
calculating CXM.

**Listing 11.1 Fortran-Like Subroutine to Calculate Eqs. (11.23)–(11.26)
for a Full Cycle. From Listing 1 in Baker (2018c)**

```
SUBROUTINE SUDOSC (X, SXM, CXM, NORD)
     REAL X(1), SXM(1), CXM(1)
     INTEGER NORD(1)
     XM = X MODULO 1
     XM2 = XM MODULO 0.5
     A = 2.0*XM2-0.5
     A = A*A
     IF (NORD = 0) THEN
            SXM = 1.0-4.0*A
            IF (XM <= 0.5) SXM = -SXM
     ELSEIF (NORD > 0) THEN
            SXM = 1.0-5.0*A+4.0*A*A
            IF (NORD = 2) THEN
                    A = XM2-0.25
                    A = A*A
                    SXM = ((-78.62897*A+0.7071068)*A+0.2629467)*A+SXM
            ENDIF
     ENDIF
     IF (XM > 0.5) SXM = -SXM
     IF ((0.25<XM) AND (XM<0.75)) THEN
            CXM = -SQRT(1-SXM*SXM)
     ELSE
            CXM = SQRT(1-SXM*SXM)
     ENDIF
RETURN
```

11.9 Constructing Full-Space Tables of Moments and Relative Amplitudes from a Few FFTs

The Fast Fourier Transform, or FFT, is linear. If $X(f)$ and $Y(f)$ are the respective
complex Fourier transforms of $x(t)$ and $y(t)$, and exist, then Eq. (11.27) holds true.

$$a * x(t) + b * y(t) \Leftrightarrow a * X(f) + b * Y(f) \qquad (11.27)$$

Take for example, the signals in the linear Eqs. (11.2) and (11.3). Let Eq. (11.3)
be N1 over S2, S3 in a common-point connection circuit, with the string signals
$n1(t)$, $-s2(t)$ and $-s3(t)$, where Eq. (11.28) shows the humbucking output equation.
This form has 3 coefficients, 1, ½ and ½. It can also be expressed in ABC form as in

Eq. (11.29). In this form it is clear that only 2 coefficients and 2 humbucking signals, $(a - b)$ and $(b - c)$ are needed to define the humbucking frequency spectra in S-U-space.

$$Vo(t) = n1(t) - [-s2(t) - s3(t)]/2 = n1(t) + [s2(t) + s3(t)]/2$$

$$\Leftrightarrow \tag{11.28}$$

$$\mathcal{V}\mathbf{o}(f) = \mathcal{N}\mathbf{1}(f) + [\mathcal{S}\mathbf{2}(f) + \mathcal{S}\mathbf{3}(f)]/2$$

where $a(t)$ is $n1(t)$, $b(t)$ is $- s2(t)$, and $c(t)$ is $- s3(t)$

$$Vo(t) = a(t) - b(t)/2 - c(t)/2 = (a(t) - b(t)) + (b(t) - c(t))/2$$
$$= s(a - b) + u(b - c), \quad s = 1, \ u = 1/2, \tag{11.29}$$
$$\Leftrightarrow$$
$$\mathcal{V}\mathbf{o}(f) = (\mathcal{A}(f) - \mathcal{B}(f)) + (\mathcal{B}(f) - \mathcal{C}(f))/2 = s(\mathcal{A} - \mathcal{B}) + u(\mathcal{B} - \mathcal{C})$$

For another example, Eq. (11.30) shows a humbucking basis vector equation, for pickup A S-up and pickups B, C and D N-up, as would happen for Fig. 11.4. This time the string signals are $-A(t)$, $B(t)$, $C(t)$ and $D(t)$. Since its vibration signal is by convention the opposite polarity of the hum signal, an S-up pickup would be connected with its minus terminal to the $+$ side of U1, and a N-Up would be connected by its plus terminal to the $-$ side of U1. Equation (11.31) shows how the Fourier transforms of the humbucking pair signals add linearly to produce the Fourier transform of the output signal, Vo. Equation (11.32) shows how the individual magnitudes of the spectral components of Vo, as determined by Eq. (11.33), are used to get the amplitude of the signal and the spectral moments, reprising similar equations back to Eq. 3.4.

$$Vo = \begin{bmatrix} s & u & v \end{bmatrix} \begin{bmatrix} 1 & -1 & 0 & 0 \\ 0 & 1 & -1 & 0 \\ 0 & 0 & 1 & -1 \end{bmatrix} \begin{bmatrix} -A \\ B \\ C \\ D \end{bmatrix}$$

$$= s(-\underline{A} - B) + u(B - C) + v(C - D) \tag{11.30}$$

$$\mathcal{V}\mathbf{o}(s, u, v) = F(Vo(s, u, v))$$
$$= sF(-A(t) - B(t)) + uF(B(t) - C(t)) + vF(C(t) - D(t)) \tag{11.31}$$
$$= s(-\mathcal{A}(f) - \mathcal{B}(f)) + u(\mathcal{B}(f) - \mathcal{C}(f)) + v(\mathcal{C}(f) - \mathcal{D}(f))$$

$V_n(f_n) = |$Amplitude of $\text{Vo}(s, u, v, \ldots)$ at frequency $f_n|, 1 \leq n \leq N$,

where $s, u, v, \ldots$ are humbucking basis vector scalars

$$\text{Amp}_V = \sum_{n=1}^{N} V_n, \quad \text{Amplitude of the signal Vo}(s, u, v, \ldots)$$

$$P_V(f_n) = \frac{V_n}{\sum_{n=1}^{N} V_n}, \quad \text{Probability density function}$$

$$\text{Mean} \cdot f = \sum_{n=1}^{N} f_n * P_V(f_n), \quad \text{mean frequency of Vo}(s, t, u) \tag{11.32}$$

$$\text{2nd} \cdot \text{moment} \cdot f = \sum_{n=1}^{N} (f_n - \text{mean} \cdot f)^2 * P_V(f_n)$$

$$\text{3rd} \cdot \text{moment} \cdot f = \sum_{n=1}^{2048} (f_n - \text{mean} \cdot f)^3 * P_V(f_n)$$

There are at least 3 forms of the frequency components of the Fourier transform; a cosine paired with a sine; a magnitude paired with a phase; and a real part paired with an imaginary part. From the form with real and imaginary parts of a frequency component $Z(f_j) = X(f_j) + iY(f_j)$, the magnitude and phase can be easily constructed, as shown in Eq. (11.33).

$$Z(f_i) = X(f_i) + jY(f_i), \quad \text{where } j^2 = -1$$

$$\text{Magnitude } Z(f_i) = |Z(f_i)| = \sqrt{X(f_i)^2 + Y(f_i)^2} \tag{11.33}$$

$$\text{Phase } Z(f_i) = \arctan\left(\frac{Y(f_i)}{X(f_i)}\right)$$

This means that however the strings can be excited to provide signals from each and every matched pickup coil being used, or from each humbucking pair, the simultaneous signals can be sampled and individually transformed into complex Fourier series. Often, the signals are sampled and digitized at high rates in sequence, so there is a finite time delay, t_0, between samples for different coils. Equation (3.20) in Brigham (1974) shows how to compensate for this, as shown in Eq. (11.34).

$$x(t - t_0) \Leftrightarrow X(f) * e^{-j2\pi f t_0}, \quad e^{-j2\pi f t_0} = \cos(2\pi f t_0) - j\sin(2\pi f t_0) \tag{11.34}$$

As a practical matter, sampling and digitizing rates can be 48 k-Samples/s or higher. For example, to obtain a frequency spectrum for 0 to 4 kHz, one must sample and digitize at 8 kS/s, which leaves room for sampling 6 signals in sequence at 48 kS/s. If an acceptable phase error is $1°$, or 0.1745 rad at 4 kHz, then the clock measuring t_0 must be accurate to $1/(360 * 4000 \text{ Hz}) = 0.694$ uS. Since it takes a few clock cycles of a microcontroller or microprocessor to mark a time, this suggests the need for a system clock of that many clock cycles times 1.44 MHz, or greater.

The linear nature of FFTs means that for J number of pickups, the FFTs of the signals of $J - 1$ number of humbucking pairs, sampled and digitized simultaneously (or corrected to simultaneity), can be recombined in any combination that equations like 11.30 and 11.31 can produce. So with $J - 1$ number of humbucking pair FFTs, from a guitar with J number of pickups, can be used to map the entire SUV-space, reconstructing FFTs for every point. The entire space can be mapped for output signal amplitude to set system gain and equalize output signal amplitudes over tone. And once a better standard for bright-warm tonality (than mean frequency) has been established, every point in the space can be mapped for tone, and ordered for output selection as a monotonic range from bright to warm and back.

11.10 Micro-controller Architecture for Humbucking Basis Vectors

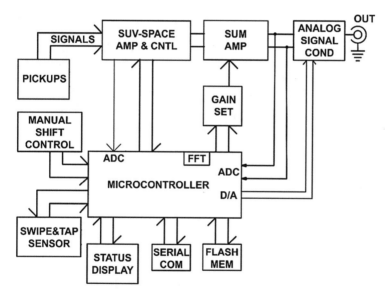

Fig. 11.27 Shows a system architecture suitable for use with a very-low-power micro-controller. It will work as well with uCs which either have trig functions or not. The PICKUPS section corresponds to Figs. 11.1–11.6 without the differential amplifiers, being matched single-coil pickups, or the coils of dual-coil humbuckers treated as single coils, with one side of the hum signal grounded on all of them. The SUV-SPACE AMP & CNTL section corresponds to Figs. 11.13, 11.14 or 11.16, but with the digital pots of Fig. 11.19. The SUM AMP and GAIN SET sections sum up the available humbucking pair signals, that have been conditioned by the vector scalars $s, u, v, \ldots$, and adjust the gain to equalize the weaker signals with the strongest. From Fig. 14 in Baker (2017c) and Fig. 21 in Baker (2018c)

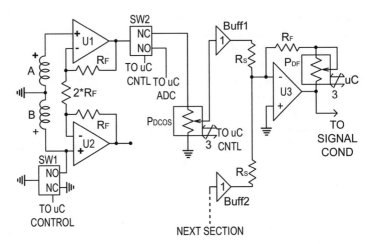

Fig. 11.28 Shows one section of a preferred embodiment of the three functional blocks, PICKUPS, SUV-SPACE AMP & CNTL, and SUM AMP in Fig. 11.27, corresponding to the A–B coil section in Fig. 11.19. The humbucking pair, A and B feed into a fully differential amplifier of gain 2, comprised of U1, U2, and the resistors R_F, R_F and $2 * R_F$. This form of differential amplifier puts virtually no load on the pickups, when the inputs are JFET or similar. For various test purposes, the solid-state 1P2T switch, SW1, shorts out pickup B on a high control signal from the uC. It is shown as the cosine section of Fig. 11.19, but a cosine section only needs connect the output of U1 to the digital pot, P_{DCOS}, simulating the SUV parameter s, which the uC would then program as a cosine pot. The solid-state 1P2T switch, SW2, on a high signal from the uC, switches the output of U1 from the digital pot to an analog-to-digital converter on the uC. This allows an FFT to be calculated from the signal (A − B). If SW1 shorts B to ground, then the A/D converter will see a signal of $2 * (A)$. This allows the FFT of pickup A alone to be calculated. Then FFT(B) = FFT (A − B) − FFT(A)/2. From Fig. 22 in Baker (2018c)

The cosine pot, P_{DCOS}, simulating the SUV parameter s, feeds into the unitary gain buffer, BUFF1, which with summing resistor R_S, and similar signals from other sections (BUFF1, R_S, . . .) sum together the humbucking pair signals, conditioned by the digital pots simulating the scalars, s, u, v, ... The feedback circuit on U3, resistor R_F and digital pot P_{DF}, provides a gain of $-(R_F + P_{DF}(set))/R_S$, as set by the uC with the 3 lines controlling P_{DF}. The output of U3 then feeds the ANALOG SIGNAL COND section in Fig. 11.27, which contains the final volume control and any tone and distortion circuits needed.

In Fig. 11.27, the output of the bypass switch, SW2 in Fig. 11.28, is shown feeding into another ADC on the uC, on its upper left, an alternative route, and another way to take FFTs and to test the circuit for faults. This kind of bypass switch can be on just one differential output signal, or on each of them, depending on the need for functional testing.

The uC in Fig. 11.27 shows 4 internal functions, one FFT section, two analog-to-digital converters, ADC, and one digital-to-analog converter, D/A. The FFT section can be a software program in the uC, an inboard or outboard Digital Signal Processor (DSP) or math co-processor that can be used to calculate FFTs, or any other functional device that serves the same purpose. The D/A output feeds inverse

FFTs to the analog output section (not shown), either as audio composites of the result of the simulation of the humbucking basis vector equation, or as a test function of various signal combinations. It allows the user to understand what the system is doing, and how. It can be embodied by a similar solid-state switch to SW1 or SW2, switching the input of the ANALOG SIGNAL COND block between the outputs of the SUM AMP and the D/A.

Ideally, the uC samples time-synced signals from all the humbucking pair signals simultaneously, performs an FFT on each one, and calculates average signal amplitudes, spectral moments and other indicia, some of which are shown in Eq. (11.32). It then uses this data to equalize the amplitudes of the entire range of possible output signals, and to arrange the tones generated into an ordered continuum of bright to warm and back. The MANUAL SHIFT CONTROL is a control input that can be embodied as anything from an up-down switch to a mouse-like roller ball to a digital display touch-swipe sensor, intended for shifting from bright to warm tones and back without the user knowing which pickups are used in what combination or humbucking basis vector sum.

So after the uC takes the FFTs of all the unmodified humbucking pair signals, via Eqs. (11.32) and (11.33), every one of the amplitudes and spectral moments, and any other measure that can be constructed from FFTs, can be calculated over the entire scalar space (s, u, v, ...). And inverse FFTs can give back representative audio signals, of every spectrum calculated, as in Eqs. (11.28) and (11.29), to check the audible order of the tones manually. With N number of pseudo-sine digital pots of 256 taps (assuming they track N pseudo-cosine pots to stay on the surface of the "potatoid"), the number of possible unique tones is 256^N. But many will be so close together as to be indistinguishable. It will still take a lot of research, experimentation and development to realize the full practical benefit of this invention.

The object of this embodiment is to allow the user to choose from and shift through a continuous gradation of tones, from bright to warm and back, automatically sequenced and controlled by the uC, so that the user never needs to know just which pickup signals are used in what combinations.

References

Baker, D. L. (2017c). Using humbucking basis vectors for generating humbucking tones from two or more matched guitar pickups, US Provisional Patent Application 62/574,705, 19 Oct 2017. Retrieved from https://www.researchgate.net/publication/335727767_Provisional_Patent_Application_62574705_Using_Humbucking_Basis_Vectors_for_Generating_Humbucking_Tones_from_Two_or_More_Matched_Guitar_Pickups

Baker, D. L. (2017d). Means and methods of controlling musical instrument vibration pickup tone and volume in STU-space, US Provisional Patent Application 62/599,452, 15 Dec 2017, continued as US Patent Application 16/156,509, 10 Oct 2018. Retrieved from https://www.researchgate.net/publication/335727919_Provisional_Patent_Application_62599452_-_

Means_and_Methods_of_Controlling_Musical_Instrument_Vibration_Pickup_Tone_and_Vol
ume_in_STU-Space

Baker, D. L. (2018c). Means and methods for obtaining humbucking tones with variable gains, US
Patent Application 16/156,509, 10 Oct 2018, published as US-2019-0057679-A1, 21 Feb 2019,
Pending. Retrieved from https://www.researchgate.net/publication/333520942_Means_and_
methods_for_obtaining_humbucking_tones_with_variable_gains_Non-Provisional_Patent_
Application_16156509_of_Donald_L_Baker_Tulsa_OK

References

Books

French, R. M. (2009). *Engineering the guitar: Theory and practice*. New York: Springer. Retrieved from https://www.springer.com/us/book/9780387743684

Patents

Anderson, T. S. (1992). Electromagnetic pickup with flexible magnetic carrier, US Patent 5,168,117, 1 Dec 1992. Retrieved from https://patents.google.com/patent/US5168117A/

Baker, D. L. (2016a). Acoustic-electric stringed instrument with improved body, electric pickup placement, pickup switching and electronic circuit. US Patent 9,401,134, July 26, 2016. Retrieved from https://patents.google.com/patent/US9401134B2/

Baker, D. L. (2019a). Humbucking switching arrangements and methods for stringed instrument pickups, US Patent 10,217,450, filed 7 June 2017, granted 26 Feb 2019. Retrieved from https://patents.google.com/patent/US10217450B2/

Baker, D. L. (2019b). Means and methods for switching odd and even numbers of matched pickups to produce all humbucking tones, US Patent Application 16/139,027, 22 Sep 2018, published as US 2019/0057678 A1, 21 Feb 2019, granted as U.S. Patent 10,380,986, 08/13/2019. Retrieved from https://patents.google.com/patent/US10380986B2/

Ball, S., et al. (2015). Musical instrument switching system, US Patent 9,196,235 B2, 24 Nov 2015. Retrieved from https://patents.google.com/patent/US9196235B2/

Ball, S., et al. (2017). Musical instrument switching system, US Patent 9,640,162 B2, 2 May 2017. Retrieved from https://patents.google.com/patent/US9640162B2/

Blucher, S. L. (1985). Transducer for stringer musical instrument. US Patent 4,501,185, February 26, 1985.

Bro, W. J. & Super, RL (2007). Maximized sound pickup switching apparatus for string instrument having a plurality of sound pickups, US Patent 7,276,657 B2, 2 Oct 2007. Retrieved from https://patents.google.com/patent/US7276657B2/

Damm, W. (2002). Single-coil electric guitar pickup with humbucking-sized housing, US Patent 6,372,976 B2, 16 Apr 2002. Retrieved from https://patents.google.com/patent/US6372976B2/

© Springer Nature Switzerland AG 2020
D. L. Baker, *Sensor Circuits and Switching for Stringed Instruments*,
https://doi.org/10.1007/978-3-030-23124-8

Fender, C. L. (1961). Electromagnetic pickup for lute-type musical instrument, US Patent 2,976,755, 28 Mar 1961. Retrieved from https://patents.google.com/patent/US2976755A/

Fender, C. L. (1966). Electric guitar incorporating improved electromagnetic pickup assembly, and improved circuit means, US Patent 3,290,424, 6 Dec 1966. Retrieved from https://patents.google.com/patent/US3290424A/

Fender, C. L. & Kaufmann, C. O. (1948). Pickup unit for stringed instruments, US Patent 2,455,575, 7 Dec 1948. Retrieved from https://patents.google.com/patent/US2455575A/

Furst. W. & Boxer, M. (2001). Sound pickup switching apparatus for a string instrument having a plurality of sound pickups, US Patent 6,316,713 B1, 12 Nov 2001. Retrieved from https://patents.google.com/patent/US6316713B1/

Hamilton, J. W. (2011). Three pickup guitar switching system with two options, US Patent 7,999,171 B1, 16 Aug 2011. Retrieved from https://patents.google.com/patent/US7999171B1/

Knapp, L. J. (1994). Electronic guitar equipped with asymmetrical humbucking electromagnetic pickup, US Patent 5,292,998, 8 Mar 1994. Retrieved from https://patents.google.com/patent/US5292998A/

Lace, M. A. (1995a). Electromagnetic musical pickup using main and auxiliary permanent magnets, US Patent 5,389,731, 14 Feb 1995. Retrieved from https://patents.google.com/patent/US5389731A/

Lace, M. A. (1995b). Electromagnetic musical pickups with central permanent magnets, US Patent 5,408,043, 18 Apr 1995. Retrieved from https://patents.google.com/patent/US5408043A/

Lesti, A. (1936). Electric translating device for musical instruments, US Patent 2,026,841, 7 Jan 1936. Retrieved from https://patents.google.com/patent/US2026841A/

Lover, S. E. (1959). Magnetic pickup for stringed musical instrument, US Patent 2,896,491, 28 July 1959. Retrieved from https://patents.google.com/patent/US2896491A/

Miessner, B. F. (1933). Method and apparatus for the production of music, US Patent 1,915,858, 27 June 1933. Retrieved from https://patents.google.com/patent/US1915858A/

Morrison, G. E. (1951). Magnetic pickup unit for guitars, US Patent 2,557,754, 19 June 1951. Retrieved from https://patents.google.com/patent/US2557754A/

Olvera, J. C. & Olvera, G. A. (2004). Electric guitar circuit control and switching module, US Patent 6,781,050 B2, 24 Aug 2004. Retrieved from https://patents.google.com/patent/US6781050B2/

Peavey, H. D. (1981). Selector switch, US Patent 4,305,320, 15 Dec 1981. Retrieved from https://patents.google.com/patent/US4305320A/

Riboloff, J. T. (1994). Guitar pickup system for selecting from multiple tonalities, US Patent 5,311,806, 17 May 1994. Retrieved from https://patents.google.com/patent/US5311806A/

Saunders, J. H. (1989). Control system with memory for electric guitars, US Patent 4,817,486, 4 Apr 1989. Retrieved from https://patents.google.com/patent/US4817486A/

Schaller, H. F. K. (1985). Magnetic pickup for stringed instruments, US Patent 4,535,668, 20 Aug 1985. Retrieved from https://patents.google.com/patent/US4535668A/

Simon, J. C. (1979). System for selection and phase control of humbucking coils in guitar pickups, US Patent 4,175,462, 27 Nov 1979. Retrieved from https://patents.google.com/patent/US4175462A/

Starr, H. W. (1987). Electric guitar pickup switching system, US Patent 4,711,149, 8 Dec 1987. Retrieved from https://patents.google.com/patent/US4711149A/

Stich, W. L. (1975). Electrical pickup for a stringed musical instrument. US Patent 3,916,751, November 4, 1975.

Thompson, P. G. (1998). Switching apparatus for electric guitar pickups, US Patent 5,763,808, 9 June 1998. Retrieved from https://patents.google.com/patent/US5763808A/

Wnorowski, T. F. (2006). Method for switching electric guitar pickups, US Patent 6,998,529 B2, 14 Feb 2006. Retrieved from https://patents.google.com/patent/US6998529B2/

Wolstein, R. J. (1992). Guitar pickup and switching apparatus, US Patent 5,136,919, 11 Aug 1992. Retrieved from https://patents.google.com/patent/US5136919A/

Patent Applications

Baker, D. L. (2016b). A switching system for paired sensors with differential outputs, especially matched single coil electromagnetic pickups in stringed instruments. US Provisional Patent Application 62/355,852, June 28, 2016. Retrieved from https://www.researchgate.net/publication

Baker, D. L. (2017a). Humbucking switching arrangements and methods for stringed instrument pickups, US Patent Application 15/616,396, filed 7 June 2017, published as US-2018-0357993-A1, 13 Dec 2018, granted as Patent US10,217,450, 26 Feb 2019. Retrieved from https://www.researchgate.net/publication/335727402_Humbucking_switching_arrangements_and_methods_for_stringed_instrument_pickups_-_NPPA_15616396

Baker, D. L. (2017b). Single-coil pickup with reversible magnet & pole sensor, US Provisional Patent Application 62/522,487, 20 June 2017. Retrieved from https://www.researchgate.net/publication/335727758_Single-Coil_Pickup_with_Reversible_Magnet_Pole_Sensor

Baker, D. L. (2017c). Using humbucking basis vectors for generating humbucking tones from two or more matched guitar pickups, US Provisional Patent Application 62/574,705, 19 Oct 2017. Retrieved from https://www.researchgate.net/publication/335727767_Provisional_Patent_Application_62574705_Using_Humbucking_Basis_Vectors_for_Generating_Humbucking_Tones_from_Two_or_More_Matched_Guitar_Pickups

Baker, D. L. (2017d). Means and methods of controlling musical instrument vibration pickup tone and volume in STU-space, US Provisional Patent Application 62/599,452, 15 Dec 2017, continued as US Patent Application 16/156,509, 10 Oct 2018. Retrieved from https://www.researchgate.net/publication/335727919_Provisional_Patent_Application_62599452_-_Means_and_Methods_of_Controlling_Musical_Instrument_Vibration_Pickup_Tone_and_Volume_in_STU-Space

Baker, D. L. (2018a). Single-coil pickup with reversible magnet & pole sensor, US Patent Application 15/917,389, 9 Mar 2018. Retrieved from https://www.researchgate.net/publication/331193192_Title_of_Invention_Single-Coil_Pickup_with_Reversible_Magnet_Pole_Sensor

Baker, D. L. (2018b). Means and methods for switching odd and even numbers of matched pickups to produce all humbucking tones, US Patent Application 16/139,027, 22 Sep 2018, published as US-2019-0057678-A1, Feb 21, 2019, granted as U.S. Patent 10,380,986, 08/13/2019. Retrieved from https://www.researchgate.net/publication/335728060_NPPA-16-139027-odd-even-HB-pu-ckts-2018-06-22

Baker, D. L. (2018c). Means and methods for obtaining humbucking tones with variable gains, US Patent Application 16/156,509, 10 Oct 2018, published as US-2019-0057679-A1, 21 Feb 2019, Pending. Retrieved from https://www.researchgate.net/publication/333520942_Means_and_methods_for_obtaining_humbucking_tones_with_variable_gains_Non-Provisional_Patent_Application_16156509_of_Donald_L_Baker_Tulsa_OK

Jacob, B. L. (2009). Programmable switch for configuring circuit topologies, US Patent Application 2009/0308233 A1, 17 Dec 2009. Retrieved from https://patents.google.com/patent/US20090308233A1/

Krozack, E., et al. (2005). Multi-mode multi-coil pickup and pickup system for stringed musical instruments, US Patent Application 2005/0150364A1, 14 July 2005. Retrieved from https://patents.google.com/patent/US20050150364A1/

Software

Steer, W. A. (2001–2016). SpecAn_3v97c.exe, Simple Audio Spectrum Analyzer v3.9, ©W.A. Steer 2001–2016. Retrieved from http://www.techmind.org/audio/specanaly.html

Ansoft Corp. (1984–2002). Maxwell, Student Version 3.1.04. Pittsburgh, PA: Ansys. Ansoft Corporation, © 1984–2002 (Acquired by Ansys, 2008). Retrieved from www.ansys.com

Standards

IEC. (2013). EC 61672-1:2013 Electroacoustics—Sound level meters—Part 1: Specifications. Retrieved from https://webstore.iec.ch/publication/5708

Index

© Springer Nature Switzerland AG 2020
D. L. Baker, *Sensor Circuits and Switching for Stringed Instruments*,
https://doi.org/10.1007/978-3-030-23124-8

Printed in the United States
by Baker & Taylor Publisher Services